Current Topics in Behavioral Neurosciences

Volume 34

About this Series

Current Topics in Behavioral Neurosciences provides critical and comprehensive discussions of the most significant areas of behavioral neuroscience research, written by leading international authorities. Each volume offers an informative and contemporary account of its subject, making it an unrivalled reference source. Titles in this series are available in both print and electronic formats.

With the development of new methodologies for brain imaging, genetic and genomic analyses, molecular engineering of mutant animals, novel routes for drug delivery, and sophisticated cross-species behavioral assessments, it is now possible to study behavior relevant to psychiatric and neurological diseases and disorders on the physiological level. The *Behavioral Neurosciences* series focuses on "translational medicine" and cutting-edge technologies. Preclinical and clinical trials for the development of new diagnostics and therapeutics as well as prevention efforts are covered whenever possible.

More information about this series at http://www.springer.com/series/7854

Suzanne Nielsen · Raimondo Bruno
Susan Schenk
Editors

Non-medical and Illicit Use of Psychoactive Drugs

Springer

Editors
Suzanne Nielsen
National Drug and Alcohol Research Centre
University of New South Wales
Sydney, NSW
Australia

Susan Schenk
School of Psychology
Victoria University of Wellington
Wellington
New Zealand

Raimondo Bruno
National Drug and Alcohol Research Centre
University of New South Wales
Sydney, NSW
Australia

and

School of Psychology
University of Tasmania
Launceston, TAS
Australia

ISSN 1866-3370 ISSN 1866-3389 (electronic)
Current Topics in Behavioral Neurosciences
ISBN 978-3-319-60014-7 ISBN 978-3-319-60016-1 (eBook)
DOI 10.1007/978-3-319-60016-1

Library of Congress Control Number: 2017949867

Cover illustration: Artistic representation of oscillatory synchrony and timing of neurons in networks by Gyorgy Buzsaki.

Printed on acid-free paper

This Springer imprint is published by Springer Nature
The registered company is Springer International Publishing AG
The registered company address is: Gewerbestrasse 11, 6330 Cham, Switzerland

Preface

New Directions in Substance Use—the Blurring of the Lines between Illicit and Therapeutic Drugs

This edited volume appears at a time of great flux for all those working in the field of psychoactive substances. For decades there has been an explicit divide between 'good' and 'bad' drugs—legal substances prescribed for therapeutic indications, and illegal substances used for intoxication. Those historical boundaries are rapidly blurring: currently North America is experiencing a self-described 'opioid crisis'(Kolodny et al. 2015), with escalating unintentional overdose deaths from extramedical use of prescription opioids (Rudd et al. 2016). At the same time, cannabis, a long-time illicit substance, is increasingly being accepted by jurisdictions internationally for use both as a medication and an intoxicant (Spithoff et al. 2015; Whiting et al. 2015). Psychedelics, often demonised as an indulgence of the counterculture, are steadily being returned to the fold of 'respectable' substances and seriously examined for therapeutic applications (Mithoefer et al. 2016; Rucker et al. 2016). The dizzying array of 'new' drugs, novel psychoactive substances, emerging at a rate of more than one entirely new substance per week between 2012 and 2015, have largely emerged from adapting candidate substances developed through the medications discovery process—from the vast annals of pharmaceutical patents and ligands developed for neuroscience—for sale in grey markets for intoxication (Mounteney et al. 2016; Sutherland et al. 2016; Zawilska and Andrzejczak 2015).

In this rapidly changing period, this edited volume, examining both the transitions of pharmaceuticals from substances for therapeutic application into desirable substances for intoxication, and transitions of 'illegal' substances back to the fold of regulated products, appears timely. The monograph follows three core themes: preclinical work with 'therapeutic' and 'illicit' substances; use of pharmaceuticals in 'non-medical' ways; and new methods for regulating previously 'illicit' substances for medical and recreational applications.

Preclinical Studies

Sanchez and Ellenbroek (2016) describe the complexities of psychopharmacological activity of antipsychotic medications and the approaches taken to testing candidate medications in the drug development process. Given the historical roots of schizophrenia as a disorder arising from dopaminergic system dysfunction (Brisch et al. 2014) it is logical that one of the common screening tests for substances was the ability to reverse the behavioural effects of dopamine stimulating substances, such as the illicit substance, amphetamine. As we have come to understand more about the complexities of schizophrenia itself (Howes et al. 2015; Khandaker and Dantzer 2016), behavioural screening tests have likewise developed. However, Sanchez and Ellenbroek (2016) highlight the curious discrepancy that all antipsychotics appear to inhibit dopaminergic activity in the very brain system that all substances of intoxication, directly or indirectly, acutely enhance —yet, despite this, there is human evidence of 'misuse' of the antipsychotic drug quetiapine (Montebello and Brett 2015). Quetiapine itself has a complex profile of pharmacological activity (Riedel et al. 2007), and it is possible that understanding the processes underlying the reinforcing processes seen in non-medical consumers may uncover new applications for this substance.

MDMA is another 'illicitly' used substance, with an unusual pharmacology. Initially pioneered in psychotherapeutic applications (Pentney 2001), the substance crossed into recreational markets, becoming the hallmark of the 1990s 'rave' culture, and remains popular today (Ort et al. 2014). One of the pioneers of MDMA research, Shuglin, noted that the drug seems to lose its 'magic' on repeated recreational use (Shulgin and Shulgin 1991). Schenk and Aronsen (2015) propose the neural underpinnings for this process: rather than simple tolerance, that neuroadaptive processes shift the drug to become a less efficacious releaser of serotonin and a more efficacious releaser of dopamine on repeated use. This has important implications for current work seeking to apply MDMA to the treatment of PTSD (Sessa and Nutt 2015): if, as seems to be the case, the drug's serotonergic empathogenic effects underpin its clinically beneficial effects (Amoroso 2015), then these studies suggest that the application of the drug in clinical contexts may be highly time-limited.

Non-medical Use of Therapeutic Substances

The non-medical use of pharmaceuticals presents a unique challenge to clinicians and regulators. Walsh and Babalonis (2016) review the development of the science covering 100 years of abuse liability testing with strong opioids. In more recent years, however, concerns with non-medical use has extended beyond the strong opioids and restricted stimulant medications (Clemow 2015): this edited volume outlines non-medical use of the weak opioid codeine, which is available without prescriptions in pharmacy (Nielsen and Van Hout 2016); as well as benzodiazepines (Nielsen 2015); and antipsychotics such as quetiapine (Montebello and Brett 2015) to which access is less stringently restricted than strong opioids.

These chapters highlight the tensions between making medications available for therapeutic use while preventing non-medical use and related harms. The complexity arises from the need to achieve balance between regulation and availability, particularly where those that require these medicines for therapeutic use are at times the same people at risk of non-medical use and harms. Even with the most highly regulated prescription drugs, the strong opioids, there are increasing rates of harms with increased utilisation, underpinning the opioid death 'epidemic' described in the United States (Han et al. 2015). Prescription monitoring frameworks for supporting quality use of medicines is often regarded as the panacea for this issue. However, the increased rates of extramedical use of weaker opioids highlights one of the limitations of prescription monitoring frameworks for strong opioids: these regulatory processes may not entirely discourage extramedical use and partially displace the challenge to less carefully regulated pharmaceuticals that carry their own unique challenges for health (Islam and McRae 2014). A clear example of this is the case of toxicity arising from chronic use of combination ibuprophen-codeine products (Frei et al. 2010).

With similar concerns raised with a variety of drugs with varying potencies and mechanisms of action, the 'bigger picture' issues relating to patient care are raised. System level changes are recommended to respond to complex medical, psychological and social conditions, recognising that medications alone are unlikely to resolve complex health conditions (Montebello and Brett 2015), (Nielsen and Van Hout 2016).

Lastly, these chapters raise the question as to how best to pragmatically respond to pharmaceutical opioids to being 'illicit' drugs and used in ways to facilitate intoxication. McLean et al. (2016) demonstrate the significant harms that can occur when individuals inject pharmaceutical products, not from the active drug itself, but from the excipients, or non-psychoactive delivery components of these pharmaceuticals such as talc, that may not dissolve when the drugs are prepared for injection. The inclusion of pharmaceuticals as a core part of the 'illicit' drug market demands both regulatory responses to reduce access and practical interventions, such as increased availability of filters along with sterile injection equipment, in order to reduce harm across the population.

New Ways of Approaching 'Illicit Drugs'

The 2016 US election saw a dramatic expansion in the numbers of US states allowing legal access to cannabis both for medical and recreational purposes, which is a trend that is also rapidly occurring internationally (Pardo 2014; Room 2014). Health bodies, the general public and politicians appear to have mixed feelings in relation to this 'liberalisation' of access to cannabis, citing concerns about impacts on mental health, road safety, and substance use disorders (Sznitman and Zolotov 2015; Watson and Mann 2016). This highlights the importance of careful evaluation of the impact of these policies. Clear evaluation of the experience following policy change will allow those jurisdictions yet to consider policy to implement optimised access policies that will support overall population health; and to provide evidence-based responses for political or emotional concerns raised in public consultations prior to implementation. Hunt and Miles (2015) demonstrate the importance of careful evaluation design for clear interpretation of the impacts of changes in cannabis policy.

There is much that we don't yet know about the impacts of bringing substances into the legal fold on the nature of existing drug markets. Wilkins et al (2017) provide evidence for a decline in natural cannabis use among police detainees in New Zealand during a period where synthetic cannabinoid receptor agonists (SCRA) were practically legal (Wilkins et al. 2013). While such a decline in illicit substance use may at one glance appear a positive trend, the net impact on public health may actually be negative—in this case given that the 'legal' SCRA products had poor manufacturing standards leading to unpredictable outcomes for consumers; the substances themselves have far greater levels of activity than natural cannabis; and the adverse effects profile, both centrally and peripherally, is far worse for SCRA than for natural cannabis (Castaneto et al. 2014; Tait et al. 2016).

Future Directions

Substances are rarely used in isolation: a critical issue for future work to tackle is the question of how to conduct research into the effects of poly-drug use. For example, among people using opioids in both clinical and illicit contexts, 33–100% also report current benzodiazepine use (Lintzeris and Nielsen 2010). Further, same day use of multiple substances including benzodiazepines and alcohol have been demonstrated among those using non-medical opioids (Peacock et al. 2016). This places increasing importance on the study of drug interactions, and how concurrent substance use affects the safety of prescribed and illicit substance use.

Rather than considering polysubstance use as solely increasing risk of harm, synergistic effects of co-administered drugs can be harnessed to increase the therapeutic potential of both substances, and reduce the doses required for either substance. Medications that are 'opioid-sparing' may enable a reduced opioid dose without loss of analgesic effect. Using opioid-sparing medications, for example paracetamol or ketamine, is one strategy to enable lower opioid dose requirements (Maund et al. 2011; Subramaniam et al. 2004). Using combinations of medications can maximise analgesic response by harnessing complementary but distinct mechanisms of action (Stoneet al. 2014). Further, use of lower doses of individual medications, through capitalising on synergistic drug effects, can improve the adverse-effect profile of each medication (Gilron et al. 2013). This is likely to be an increasingly important area of study given the limitations in effectiveness of single substances, and the harms from escalating opioid doses.

Together, the diverse research threads collected in this edited volume demonstrate that the simplistic binary categorisation of drugs into 'good' vs. 'bad'; 'therapeutic' vs. 'intoxicant' camps may no longer be meaningful. This calls on researchers from both traditions to carefully consider both 'sides' in their work—for pharmaceutical researchers to consider abuse liability and the potential health harms of their products (Alexanderet al. 2014; Webster and Fine 2010); and for illicit drug use researchers to also consider the potential benefits of substances for consumers (Moore 2008). Such a revolution in approach will require regulatory frameworks to evolve rapidly alongside the changes in the science.

Sydney, Australia — Suzanne Nielsen
Sydney/Launceston, Australia — Raimondo Bruno
Wellington, New Zealand — Susan Schenk

References

Alexander L, Mannion RO, Weingarten B, Fanelli RJ, and Stiles GL (2014). Development and impact of prescription opioid abuse deterrent formulation technologies. Drug and Alcohol Dependence, 138, 1–6. doi:10.1016/j.drugalcdep.2014.02.006

Amoroso T (2015). The psychopharmacology of +/-3,4 methylenedioxymethamphetamine and its role in the treatment of posttraumatic stress disorder. J Psychoactive Drugs, 47(5), 337–344. doi:10.1080/02791072.2015.1094156

Brisch R, Saniotis A, Wolf R, Bielau H, Bernstein HG, Steiner J, Bernhard B, Katharina B, Zbigniew J, Jaliya K, Maciej H, Gos T (2014). The role of dopamine in schizophrenia from a neurobiological and evolutionary perspective: old fashioned, but still in vogue. Front Psychiatry, 5:47. doi:10.3389/fpsyt.2014.00047

Castaneto MS, Gorelick DA, Desrosiers NA, Hartman RL, Pirard S, and Huestis MA (2014). Synthetic cannabinoids: epidemiology, pharmacodynamics, and clinical implications. Drug and Alcohol Dependence, 144, 12–41. doi:10.1016/j.drugalcdep.2014.08.005

Frei MY, Nielsen S, Dobbin MD, and Tobin CL (2010). Serious morbidity associated with misuse of over-the-counter codeine-ibuprofen analgesics: a series of 27 cases. Med J Aust, 193(5), 294–296

Gilron I, Jensen TS, and Dickenson AH (2013). Combination pharmacotherapy for management of chronic pain: from bench to bedside. The Lancet Neurology, 12(11), 1084–1095

Han B, Compton WM, Jones CM, and Cai R (2015). Nonmedical prescription opioid use and use disorders among adults aged 18 through 64 years in the united states, 2003–2013. JAMA, 314 (14), 1468–1478. doi:10.1001/jama.2015.11859

Howes O, McCutcheon R, and Stone J (2015). Glutamate and dopamine in schizophrenia: an update for the 21st century. J Psychopharmacol, 29(2), 97–115. doi:10.1177/0269881114563634

Islam MM, and McRae IS (2014). An inevitable wave of prescription drug monitoring programs in the context of prescription opioids: pros, cons and tensions. BMC Pharmacol Toxicol, 15:46. doi:10.1186/2050-6511-15-46

Khandaker GM, and Dantzer R (2016). Is there a role for immune-to-brain communication in schizophrenia? Psychopharmacol, 233(9), 1559–1573. doi:10.1007/s00213-015-3975-1

Kolodny A, Courtwright DT, Hwang CS, Kreiner P, Eadie JL, Clark TW, and Alexander GC (2015). The prescription opioid and heroin crisis: a public health approach to an epidemic of addiction. Annu Rev Public Health, 36, 559–574. doi:10.1146/annurev-publhealth-031914-122957

Lintzeris N, and Nielsen S (2010). Benzodiazepines, Methadone and buprenorphine: interactions and clinical management. Am J Addict, 19(1), 59–72

Maund E, McDaid C, Rice S, Wright K, Jenkins B, and Woolacott N (2011). Paracetamol and selective and non-selective non-steroidal anti-inflammatory drugs for the reduction in morphine-related side-effects after major surgery: a systematic review. Br J Anaesth, 106(3), 292–297. doi:10.1093/bja/aeq406

Mithoefer MC, Grob CS, and Brewerton TD (2016). Novel psychopharmacological therapies for psychiatric disorders: psilocybin and MDMA. Lancet Psychiatry, 3(5), 481–488. doi:10.1016/s2215-0366(15)00576-3

Montebello ME, and Brett J (2015). Misuse and associated harms of quetiapine and other atypical antipsychotics. Curr Top Behav Neurosci. doi:10.1007/7854_2015_424

Moore D (2008). Erasing pleasure from public discourse on illicit drugs: on the creation and reproduction of an absence. Int J Drug Policy, 19(5), 353–358. doi:10.1016/j.drugpo.2007.07.004

Mounteney J, Griffiths P, Sedefov R, Noor A, Vicente J, and Simon R (2016). The drug situation in Europe: an overview of data available on illicit drugs and new psychoactive substances from European monitoring in 2015. Addict, 111(1), 34–48. doi:10.1111/add.13056

Nielsen S (2015). Benzodiazepines. Curr Top Behav Neurosci. doi:10.1007/7854_2015_425

Nielsen S, and Van Hout MC (2016). Over-the-counter codeine-from therapeutic use to dependence, and the grey areas in between. Curr Top Behav Neurosci. doi:10.1007/7854_2015_422

Ort C, van Nuijs ALN, Berset J-D, Bijlsma L, Castiglioni S, Covaci A, Voogt P, Emke E, Fatta-Kassinos D, Griffiths P, Hernández F, González-Mariño I, Grabic R, Kasprzyk-Hordern B, Mastroianni N, Meierjohann A, Nefau T, Östman M, Pico Y, Racamonde I, Reid M, Slobodnik J, Terzic S, Thomaidis N, Thomas KV (2014). Spatial differences and temporal changes in illicit drug use in Europe quantified by wastewater analysis. Addict, 109(8), 1338–1352. doi:10.1111/add.12570

Pardo B (2014). Cannabis policy reforms in the Americas: a comparative analysis of Colorado, Washington, and Uruguay. Int J Drug Policy, 25(4), 727–735. doi:10.1016/j.drugpo.2014.05.010

Peacock A, Bruno R, Larance B, Lintzeris N, Nielsen S, Ali R, Dobbins T, Degenhardt, L. (2016). Same-day use of opioids and other central nervous system depressants amongst people who tamper with pharmaceutical opioids: a retrospective 7-day diary study. Drug Alcohol Depend, 166, 125–133. doi:10.1016/j.drugalcdep.2016.07.003

Pentney AR (2001). An exploration of the history and controversies surrounding MDMA and MDA. J Psychoactive Drugs, 33(3), 213–221. doi:10.1080/02791072.2001.10400568

Riedel M, Müller N, Strassnig M, Spellmann I, Severus E, and Möller H-J (2007). Quetiapine in the treatment of schizophrenia and related disorders. Neuropsychiatric Dis Treat, 3(2), 219–235

Room R (2014). Legalizing a market for cannabis for pleasure: colorado, washington, uruguay and beyond. Addict, 109(3), 345–351. doi:10.1111/add.12355

Rucker JJ, Jelen LA, Flynn S, Frowde KD, and Young AH (2016). Psychedelics in the treatment of unipolar mood disorders: a systematic review. J Psychopharmacol. doi:10.1177/0269881116679368

Rudd RA, Aleshire N, Zibbell JE, and Matthew GR (2016). Increases in drug and opioid overdose deaths—United States, 2000–2014. Am J Transplant, 16(4), 1323–1327. doi:10.1111/ajt.13776

Sessa B, and Nutt D (2015). Making a medicine out of MDMA. Br J Psychiatry, 206(1), 4–6. doi:10.1192/bjp.bp.114.152751

Shulgin A, and Shulgin A (1991). *PiHKAL: A chemical love story*. California: Transform Press

Spithoff S, Emerson B, and Spithoff A (2015). Cannabis legalization: adhering to public health best practice. Can Med Assoc J, 187(16), 1211–1216. doi:10.1503/cmaj.150657

Stone LS, German JP, Kitto KF, Fairbanks CA, and Wilcox GL (2014). Morphine and clonidine combination therapy improves therapeutic window in mice: synergy in antinociceptive but not in sedative or cardiovascular effects. PLoS One, 9(10), e109903. doi:10.1371/journal.pone.0109903

Subramaniam K, Subramaniam B, and Steinbrook RA (2004). Ketamine as adjuvant analgesic to opioids: a quantitative and qualitative systematic review. Anesth Analg, 99(2), 482–495. doi:10.1213/01.ane.0000118109.12855.07

Sutherland R, Peacock A, Whittaker E, Roxburgh A, Lenton S, Matthews A, Butler K, Nelson M, Burns L, Bruno, R. (2016). New psychoactive substance use among regular psychostimulant users in Australia, 2010–2015. Drug Alcohol Depend, 161, 110–118. doi:10.1016/j.drugalcdep.2016.01.024

Sznitman SR, and Zolotov Y (2015). Cannabis for therapeutic purposes and public health and safety: a systematic and critical review. Int J Drug Policy, 26(1), 20–29. doi:10.1016/j.drugpo.2014.09.005

Tait RJ, Caldicott D, Mountain D, Hill SL, and Lenton S (2016). A systematic review of adverse events arising from the use of synthetic cannabinoids and their associated treatment. Clin Toxicol, 54(1), 1–13. doi:10.3109/15563650.2015.1110590

Watson TM, and Mann RE (2016). International approaches to driving under the influence of cannabis: a review of evidence on impact. Drug and Alcohol Dependence, 169, 148–155. doi:10.1016/j.drugalcdep.2016.10.023

Webster LR, and Fine PG. (2010). Approaches to improve pain relief while minimizing opioid abuse liability. J Pain, 11(7), 602–611. doi:10.1016/j.jpain.2010.02.008

Whiting PF, Wolff RF, Deshpande S, Di Nisio M, Duffy S, Hernandez AV, Keurentjes JC, Lang S, Misso K, Ryder S, Schmidlkofer S, Westwood M, Kleijnen J (2015). Cannabinoids for medical use: a systematic review and meta-analysis. JAMA, 313(24), 2456–2473. doi:10.1001/jama.2015.6358

Wilkins C, Sheridan J, Adams P, Russell B, Ram S, and Newcombe D (2013). The new psychoactive substances regime in New Zealand: a different approach to regulation. J Psychopharmacol, 27(7), 584–589. doi:10.1177/0269881113491441

Zawilska JB, and Andrzejczak D (2015). Next generation of novel psychoactive substances on the horizon—A complex problem to face. Drug and Alcohol Dependence, 157, 1–17. doi:10.1016/j.drugalcdep.2015.09.030

Contents

Preclinical Effects of Antipsychotic Drugs

Jenny Paola Berrío Sánchez and Bart A. Ellenbroek

Abstract Antipsychotic drugs have been the drugs of choice for the treatment of schizophrenia ever since the introduction of chlorpromazine in the early 1950s of the last century. Since then, about 60 different antipsychotics have been introduced. Although pharmacologically these drugs show large differences, in terms of potency, duration of action and selectivity, all antipsychotics appear to reduce the positive symptoms of schizophrenia, while having little or no effect on the negative symptoms or the cognitive deficits. The only apparent exception is clozapine, which is also effective in therapy-resistant patients. On the other hand, antipsychotics induce significant side effects as well, including neurological, behavioural and metabolic side effects. In the present paper, we will discuss the preclinical pharmacology of the current antipsychotic drugs focussing both on the therapeutic and on side effects of these drugs.

Keywords First-generation · Second-generation · Schizophrenia · Bipolar disorder · Side-effects · Dopamine

Contents

J.P.B. Sánchez · B.A. Ellenbroek (✉)
School of Psychology, Victoria University of Wellington,
PO Box 600, Wellington, New Zealand
e-mail: bart.ellenbroek@vuw.ac.nz

Curr Topics Behav Neurosci (2017) 34:1–16
DOI 10.1007/7854_2016_447
Published Online: 30 June 2016

1 Introduction

The 1950s was a period of unprecedented breakthroughs in the pharmacological treatment of major psychiatric disorders. In 1952, the antidepressant activity of iproniazid was detected; in 1953, chlorpromazine was found to selectively reduce psychotic symptoms in patients with schizophrenia; in 1955, imipramine's antidepressant activity was identified; and finally, in 1957, chlordiazepoxide was found to have powerful anxiolytic effects. Interestingly, all these drugs were found serendipitously. Iproniazid was a known antituberculosis drug, chlorpromazine was thought to be a sedative antihistaminic, imipramine (due to its structural similarity to chlorpromazine) was expected to have antipsychotic effects, and chlordiazepoxide was thought to be a chemical dye as it was synthesized from the quinazoline 3-oxide class of dyes. It was not until the 1960s and 1970s, when many additional effective drugs were synthesized and marketed, that a more systematic investigation into the mechanism(s) of action of these psychopharmacological drugs was undertaken. Shortly after the introduction of chlorpromazine and several related compounds, Delay and Deniker coined the term "neuroleptic drugs" for this group of compounds (from the Greek "that which seizes the nerves"). Although it is often thought that the term refers to the neurological side effects, Delay and Deniker considered these incidental and unrelated to the therapeutic effects. Rather, they emphasized the sedative and emotional effects (Moncrieff 2013). Only after the work of Steck in Switzerland (Steck 1954) and especially of Haase in Germany (Haase 1954), the neurological symptoms (usually referred to as extrapyramidal side effects) were considered to be intimately related to the therapeutic effects and the Haase's concept of the neuroleptic threshold was introduced (Haase et al. 1974).

The idea behind the neuroleptic threshold was that subtle changes in motor symptoms (especially rigidity) were essential for obtaining a therapeutic effect. Haase used a specific writing test to determine, during his dose-finding studies, when these symptoms occurred (indicated by micrographia). This dose was considered the neuroleptic threshold dose and was maintained in each patient until remission of symptoms occurred. However, several multicentre studies in the USA during the early 1960s had already shown that different antipsychotics induced different degrees of motor symptoms, and with the introduction of clozapine that induces virtually no extrapyramidal side effects, the concept of "atypical neuroleptics" (or atypical antipsychotics as they are now generally referred to) was born. Although this terminology is still in use, it is now more customary to refer to these drugs as second-generation antipsychotics, as opposed to the (older) first-generation antipsychotics. The reasons for this have been discussed elsewhere (Ellenbroek and Cesura 2015), but are principally related to the realization that the atypical antipsychotics do not really represent a homogenous class of drugs, but contain many diverse members. Table 1 lists the different first- and second-generation antipsychotics that have been approved for the treatment of schizophrenia. Some of these (especially the second-generation drugs) have also

Table 1 FDA approved first- and second-generation antipsychotic drugs for the treatment of schizophrenia

First-generation antipsychotics	Second-generation antipsychotics
Chlorpromazine	Aripiprazole (2002)
Fluphenazine	Asenapine (2009)
Haloperidol	Brexpiprazole (2015)
Loxapine	Cariprazine (2015)
Perphenazine	Clozapine (1990)
Prochlorperazine	Iloperidone (2009)
Thiothixene	Lurasidone (2010)
Thioridazine	Olanzapine (1996)
Trifluoperazine	Paliperidone (2006)
	Quetiapine (1997)
	Risperidone (1996)
	Ziprasidone (2001)

The number in brackets represents the year of approval by the FDA

been approved for the treatment of other indications, such as Tourette syndrome, bipolar disorder and treatment-resistant depression.

As discussed at length in several other papers although there are clear differences between the different antipsychotic drugs, these are mainly related to the side effect profile (Ellenbroek 2012; Ellenbroek and Ghiabi 2015, Ellenbroek and Cesura 2015). With respect to the therapeutic profile, all approved first- and second-generation antipsychotic drugs improve positive symptoms, but have little or no effect on negative and cognitive symptoms (Leucht et al. 2013; Fusar-Poli et al. 2015). It is now well established that this common drug effect is also mediated via a single common drug action, namely blockade of the dopamine D2 receptor. As extensively reviewed elsewhere (Madras 2013; Ellenbroek and Cesura 2015), this discovery spans several decades, beginning with the pioneering biochemical and behavioural work in animals (Carlsson et al. 1963; van Rossum 1966) and culminating in the in vivo measurement of dopamine receptor occupancy in patients (Farde et al. 1988). This latter study showed that a blockade of about 60–70 % of dopamine D2 receptors is necessary to obtain a therapeutically satisfactory reduction in positive symptoms. In addition, studies have shown that when the blockade reaches 80–85 %, clinically significant extrapyramidal side effects are observed (Nord and Farde 2011). Although these data seem directly opposite to Haase's neuroleptic threshold, they do indicate that the therapeutic width is, in general, quite small. Clozapine seems to be the only exception to this general rule, as its therapeutic width appears very large, and a reduction in symptoms can already be observed at occupancy levels around 30–40 % (Farde et al. 1988).

Thus, while the therapeutic effects of most antipsychotic drugs are very similar, substantial differences have been found with respect to the side effects. On the one hand, first-generation drugs such as haloperidol and zotepine and second-generation drugs such as lurasidone and risperidone induce significantly more extrapyramidal side effects than clozapine, olanzapine and quetiapine (Kahn et al. 2008; Leucht et al. 2013). On the other hand, metabolic side effects (especially weight gain) are much more prevalent with clozapine, olanzapine and chlorpromazine and much less with haloperidol, ziprasidone and aripiprazole (Leucht et al. 2013). Finally, some recent papers have suggested to some antipsychotics, most notably quetiapine may even be abused by some individuals (see Montebello 2016).

Although these data clearly illustrate the important differences between antipsychotic drugs, they also illustrate that there is no clear subdivision between first- and second-generation antipsychotics. In summary, the clinical data appear to indicate that virtually all antipsychotics selectively improve the positive symptoms of schizophrenia by blockading the dopamine D2 receptor. In addition, virtually all antipsychotics induce a number of different side effects, some of which are directly related to the blockade of dopamine D2 receptors (such as the extrapyramidal side effects), while others are related to other actions of antipsychotics. More importantly, the clinical results show that there is clear room for improvement. With the possible exception of clozapine, none of the antipsychotic drugs significantly improve negative or cognitive symptoms and all antipsychotics have substantial side effects. Although clozapine has a superior clinical profile, its usefulness is hampered by the occurrence of agranulocytosis, a potentially lethal reduction of neutrophils (De Fazio et al. 2015). Nonetheless, clozapine's usefulness in therapy-resistant patients coupled with its exceptionally low rate of occupancy suggests that it acts (at least in part) via a novel neurobiological mechanism. Therefore, the goal (and indeed the Holy Grail) of preclinical antipsychotic research is to develop a drug similar to clozapine, which targets the positive, negative and cognitive symptoms of schizophrenia while inducing only mild side effects. Since the drug action of clozapine is extremely complex (see below), most studies have used a more pragmatic behavioural approach using screening tests. Although these tests have been helpful in the past, they present their own limitations. As such, we will need novel simulation models for schizophrenia before we can really expect a breakthrough in therapy. Nonetheless, these screening models give us insight into the preclinical pharmacology of antipsychotic drugs. In addition, several other preclinical paradigms have been developed for assessing additional effects of antipsychotics drugs, focussing more specifically on the different side effects. In the remainder of this paper, we will discuss these different preclinical models.

2 The Binding Profile of Antipsychotics

Before discussing the behavioural pharmacology of antipsychotic drugs, it is important to discuss briefly how these drugs affect brain processes. As mentioned above, all antipsychotics block dopamine D2 receptors (with aripiprazole acting as a partial agonist), but most antipsychotics also block a myriad of other receptors. As an example, Fig. 1 shows the binding profiles of the first-generation antipsychotics haloperidol and chlorpromazine in comparison with clozapine, risperidone and olanzapine as examples of second-generation antipsychotics. The figure illustrates several important points. First of all, and not surprisingly given the discussion above, all antipsychotics bind to the dopamine D2 receptor. Secondly, none of these antipsychotics are very selective, most having appreciable affinity for several other receptors, most notably the α1A and the 5-HT2A receptors. Thirdly, all antipsychotics show very different profiles. Thus, whereas haloperidol has the highest affinity for the D2 receptor, clozapine and chlorpromazine bind strongly to the α1A receptor, and olanzapine and risperidone show the greatest affinity for the 5-HT2A receptor. Finally, the second-generation antipsychotics, especially clozapine and olanzapine, show appreciable affinity for a large number of different receptors. Although Fig. 1 only shows a subgroup of first- and second-generation antipsychotics, with the exception of the group of the benzamides, the conclusions are not substantially different. The benzamides (including such drugs as the first-generation sulpiride and the second-generation amisulpride) show a relatively high selectivity for the dopamine receptors (mainly the D2 and D3 receptor), admitting some also show affinity for several 5-HT receptors (Ellenbroek and Cesura 2015).

Several theories have been proposed to explain the side effect profile of antipsychotics, mainly focussing on the presence or (relative) absence of extrapyramidal side effects. One of the earliest theories was *the muscarinic acetylcholinergic receptor theory* (Snyder et al. 1974). This was based on the clinical practice of treating extrapyramidal side effects with muscarinic antagonists and the fact that clozapine also has affinity for muscarinic receptors. However, the majority of the other antipsychotics (counting several second-generation ones) do not block the muscarinic receptors. A more influential theory was put forward by Herbert Meltzer and colleagues and became known as the *DA/5-HT2A receptor balance theory*. The main pillar of this theory is that antipsychotic drugs which

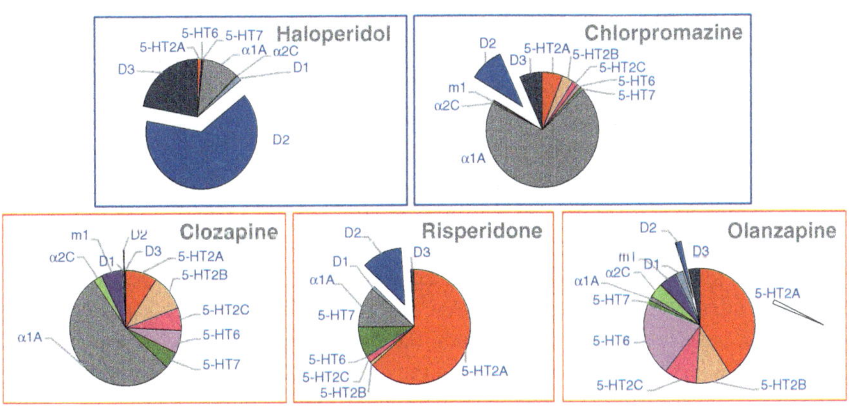

Fig. 1 The binding profile of first-generation antipsychotic (haloperidol and chlorpromazine) and second-generation antipsychotic (clozapine, risperidone and olanzapine). Data are taken from the NIMH sponsored database (see http://pdspdb.unc.edu/pdspWeb/)

block the 5-H2A receptors more strongly than the D2 receptor would induce less extrapyramidal side effects than drugs that show the reverse profile (Meltzer et al. 1989). Although this theory had quite some merit and has prevailed till today, it contains several clear limitations. First of all, it does not explain the profile of amisulpride, which induces less extrapyramidal side effects even though its affinity for the D2 receptors is significantly higher than that for the 5-HT2A receptor. What is more, it does not seem to hold up to more recent evaluations of the clinical effects of first- and second-generation antipsychotics. In the extensive meta-analysis by Stephan Leucht and colleagues, risperidone, zotepine and lurasidone were among the drugs that showed the highest incidence of extrapyramidal side effects, yet they show a strong blockade of 5-HT2A receptors (Leysen 2000; Ishibashi et al. 2010). Many other theories have emerged, but none has been proven universally applicable. Then, it seems likely that the differences between different antipsychotics cannot be explained by a single overarching theory.

3 Behavioural Test for the Therapeutic Effects of Antipsychotics

Most of the behavioural effects of antipsychotics have focussed on mimicking the therapeutic effects these drugs have on patients with schizophrenia. In fact, the vast majority of antipsychotics were developed using these so-called screening tests. Screening tests, sometimes referred to as models with pharmacological isomorphism (Matthysse 1986; Ellenbroek 1993), are based on the premise that similarities between drugs are comparable across species. This does not mean that drugs have the same effects across species, rather that when two drugs have a similar effect in one species, they are likely to have the same effect in another species. Thus, the test generally compares a new drug with an existing well-defined drug referred to as the "gold standard". In screening tests for antipsychotics, the gold standard is often chlorpromazine or haloperidol, and in some instances, two gold standards are used, such as haloperidol and clozapine (Ellenbroek et al. 1987).

Paul Janssen and his colleagues were among the first to use screening models to compare different antipsychotics. In a series of papers, they compared a large number of antipsychotics through different behavioural tests with the aim of predicting antipsychotic activity (Janssen et al. 1965, 1966, 1967). By using a large number of different drugs from various different chemical classes, they were able to assess the existence of false negatives, i.e. drugs that are effective in the clinic but fail to be effective in the model. In time, a set of additional validation criteria were developed (see Table 2). These include the evaluation of false positive (i.e. drugs that are effective in the model but not in the clinic) and several other criteria that are specifically based on the pharmacodynamics of antipsychotics in clinical practice and that aimed at addressing crucial aspects of screening tests.

Table 2 Validation criteria for screening tests for antipsychotics

Criteria
1. The number of false positives should be as low as possible
2. The number of false negatives should be as low as possible
3. There should be a reasonable correlation between the potencies in the test and the clinic
4. Anticholinergic drugs should not reduce the effect in the test
5. Chronic treatment should not reduce the effect in the test

Although there is convincing evidence that the principle of pharmacological isomorphism between species is valid, drugs invariably induce a large variety of symptoms in humans, beneficial as well as unwanted side effects. Indeed, as discussed above, virtually all antipsychotics induce extrapyramidal side effects and both the therapeutic and the extrapyramidal side effects are largely mediated by a blockade of dopamine receptors. Fortunately, a close inspection of the clinical literature shows that there are two important pharmacological differences between the antipsychotic effects and the extrapyramidal side effects. One, the latter but not the former are sensitive to anticholinergic drugs, and two, most of the extrapyramidal side effects show tolerance upon chronic treatment, while the antipsychotic effects do not (in fact, they often get stronger with repeated administration).

Many different screening tests have been developed for schizophrenia (Ellenbroek 2015); Table 3 lists the most important ones, evaluated against the criteria from Table 2. Probably, the most used test is the conditioned avoidance response (CAR) (Arnt 1982; Wadenberg and Hicks 1999). In this test, animals are trained in a two-compartment shuttle box to avoid an electric shock when a specific stimulus precedes it. The test can lead to three different outcomes: avoidance (animals move when the stimulus comes on and successfully avoids the shock); escape (animals move when the shock comes on) and failure (animals do not move at all). Once animals have successfully acquired the CAR (usually defined as 85 % correct avoidances), they are treated with an antipsychotic. As antipsychotics affect

Table 3 Several important screening tests for antipsychotic medication and their validation status with respect to the criteria of Table 2

Test	1	2	3	4	5
Conditioned avoidance response	−	+	+	−	+
Paw test	+	+	+	+	Ø
Catalepsy	−	−	−	−	Ø
Dopamine agonist-induced LMA	−	Ø	+	Ø	Ø
Dopamine agonist-induced stereotypy	−	−	−	−	−
Dopamine agonist-induced prepulse inhibition	−	+	+	−	+
NMDA antagonist-induced prepulse inhibition	−	−	−	Ø	+

\+ indicates the criterion is largely fulfilled
− indicates the criterion is violated
Ø indicates insufficient or contradictory data

motor behaviour and the correct performance requires the animals to jump, a dose–response curve is essential. For most antipsychotics, low doses reduce avoidance without affecting escape, while higher doses reduce both avoidance and escape; this has been interpreted as inducing motor side effects. Nonetheless, the assessment reveals that quite a number of false positives exist and the CAR also does not fare well with criteria 4 and 5. A number of years ago, we developed another screening test, the paw test (Ellenbroek et al. 1987). In this test, rats were injected with a drug and placed on a platform with four holes. Each of the four limbs was placed within one hole, and the time to withdraw one forelimb and one hindlimb was measured. In contrast to the other tests, the paw test was assessed using two gold standards: haloperidol and clozapine. While haloperidol affected forelimb and hindlimb equally, clozapine only affected the hindlimb retraction time. Subsequent studies showed that the paw test fares reasonably well with all the criteria (Ellenbroek and Cools 1988). Most second-generation antipsychotics are more effective on the hindlimb retraction time, while most first-generation antipsychotics affect both forelimb and hindlimb to a similar degree. Moreover, anticholinergic and chronic treatment reduces the effect on forelimb but not hindlimb retraction time. Thus, these data indicate that the neurobiological mechanisms underlying the forelimb and hindlimb retraction time are different. Perhaps more intriguingly, the neurobiological mechanisms underlying the effects of haloperidol and clozapine on the hindlimb retraction time are different. Thus, whereas the effects of haloperidol could be reversed by a dopamine D2 agonist, the effect of clozapine was sensitive to a D1 agonist (Ellenbroek et al. 1991). Likewise, while the 5-HT1A agonist 8-OHDPAT increased the hindlimb retraction time of clozapine, it decreased that of haloperidol. Finally, the reverse situation was found with the 5-HT2 agonist DOI:an inhibition of clozapine's and a potentiation of haloperidol's effect on hindlimb retraction time (Ellenbroek et al. 1994). Together, these data are reminiscent of the clinical data showing that clozapine can be effective in haloperidol-resistant patients and hence likely produce its therapeutic effect via a different mechanism.

Given the intimate relationship between schizophrenia, antipsychotic drugs and dopamine, several different screening tests have been based on the reversal of dopamine agonist-induced behaviour such as hyperactivity and stereotypy (Janssen et al. 1967; Thomas and Handley 1978; Niemegeers and Janssen 1979; Arnt 1983, 2000). On theoretical grounds, both were originally thought to be more directly related to schizophrenia than the other screening tests (and therefore more simulation type models). In the case of amphetamine-induced hyperactivity, the rationale was twofold: first of all, amphetamine is known to induce psychotic like symptoms, and secondly, the amphetamine-induced hyperactivity is due to an increased dopamine release in the nucleus accumbens, which was thought to accompany psychotic symptoms in patients. However, more recent studies have shown that, although enhanced dopamine release is present in patients with schizophrenia, this is localized in the anterodorsal part of the caudate nucleus rather than the nucleus accumbens (Kegeles et al. 2010). The rationale behind the apomorphine-induced stereotypy is based on the findings that stereotyped behaviour is common in patients with schizophrenia, both in the motor and in cognitive domains. Since all antipsychotics block the dopamine D2 receptor, it is not surprising that all are effective in both models, although several of the second-generation antipsychotics (especially clozapine and several benzamides) appear much less effective in blocking stereotypy (Ogren et al. 1984, 1986). However, many other drugs also inhibit the behavioural effects of amphetamine and apomorphine, thus leading to many false negatives. Likewise, anticholinergic drugs have been found to reverse the effects of antipsychotics, severely limiting the predictive validity of both models. A last dopamine-related model, which also has been thought to be more related to the simulation models, is the apomorphine-induced disruption of prepulse inhibition. Prepulse inhibition refers to a reduction in the startle response by a weaker stimulus immediately preceding the startle stimulus (Graham 1975). Although startle and prepulse inhibition can be measured in several different domains, most studies have focussed on the inhibition of the acoustic startle response. One of the main advantages of prepulse inhibition is that it occurs in virtually every species and can be measured using (more or less) identical techniques in both animals and humans, which makes it a perfect candidate for translational research. Moreover, deficits in prepulse inhibition are among the most replicated findings in patients with schizophrenia (Braff et al. 1978; Braff and Geyer 1990; Braff 2015). Consequently, the apomorphine-induced disruption of prepulse inhibition was originally considered (as the other dopamine-related models) more a simulation model than a screening test, especially as all antipsychotics reversed the apomorphine-induced prepulse inhibition (Mansbach et al. 1988; Geyer et al. 2001). However, there are two reasons why the apomorphine-induced prepulse inhibition does not qualify as a true simulation model. First of all, schizophrenia is a chronic illness with a strong neurodevelopmental component and hence cannot be mimicked by a single drug injection. Secondly, and more importantly, the disruption in prepulse inhibition in schizophrenia is also seen in patients on active antipsychotic treatment, suggesting that, at best, antipsychotics only have a limited effect on prepulse inhibition. Unfortunately, longitudinal studies are limited and inconclusive. Some have suggested that first-generation antipsychotics have no effect on the deficit in prepulse inhibition, while second-generation antipsychotics have an ameliorating effect (Meincke et al. 2004; Quednow et al. 2006; Aggernaes et al. 2010); others found no improvement after 8 weeks of treatment (Mackeprang et al. 2002; During et al. 2014). Although this significantly undermines the validity of apomorphine-induced prepulse inhibition as a simulation model, it is important to note that the requirements for screening tests are different. As long as the criteria from Table 2 are fulfilled, the test is valid. Since screening models do not need to encompass symptoms of schizophrenia (and indeed rarely do), it is irrelevant whether the parameter is actually affected by antipsychotics in the clinic. In this respect, as a screening model, disruption of prepulse inhibition induced by the NMDA antagonists (including ketamine, phencyclidine and MK-801) has less predictive validity as many different antipsychotics fail to reverse such disruptions (Geyer et al. 2001).

4 Antipsychotic Drugs and Drug Discrimination

In line with the clinical findings that there are not only similarities but also differences between antipsychotics, drug discrimination studies have been used to assess the degree of (dis)similarity. Drug discrimination is based on the principle that animals (including humans) can distinguish the interoceptive cue of a drug as long as the drug acts in the central nervous system. This is usually assessed in operant chambers where animals are trained to press one lever when they are treated with a specific drug and another one when they are treated with a vehicle. Once the animals have successfully acquired this, generalization studies are performed in which a novel drug is given. If the drug induces a similar interoceptive cue, the animal will press the drug-paired lever, and if not, it will press the vehicle-paired lever. Hence, this technique allows for a detailed analysis of the similarity between different antipsychotic drugs. The first studies using chlorpromazine showed a generalization for some phenothiazines (such as perphenazine and prothipendyl) but not others such as prochlorperazine or the tricyclic antidepressant imipramine (Stewart 1962).

Subsequent studies were successful in training rats to discriminate other antipsychotics, including haloperidol and clozapine (Porter and Prus 2009). Given the aforementioned unique profile of clozapine, much research has been devoted to understanding its discriminative cue. One of the unique components of clozapine (see Fig. 1) is its strong anticholinergic activity, and indeed, some studies have indicated that muscarinic antagonists such as atropine (Nielsen 1988) and scopolamine (Kelley and Porter 1997) generalize to the clozapine cue. Moreover, the cholinergic agonist oxotremorine dose-dependently blocks the clozapine cue (Nielsen 1988). More detailed pharmacological analysis showed that it is the M1, rather than the M2 muscarinic receptor, that predominantly mediates the discriminative cue (Porter and Prus 2009). However, in line with its complex pharmacological profile, there is evidence that clozapine's discriminative cue is a compound cue involving other receptors as well (Porter and Prus 2009). Other second-generation antipsychotics successfully used as training drugs include olanzapine, quetiapine and ziprasidone. Most studies (though not all) have shown that second-generation (including clozapine) but not first-generation antipsychotics fully substitute for either of these three training drugs (Porter and Prus 2009).

Overall, these behavioural data add support for the clinical results that while there are clear similarities between antipsychotic drugs, subtle differences exist as well.

5 Behavioural Tests for the Side Effect of Antipsychotics

As referred to above, especially with regard to the side effects, antipsychotics differ significantly, with extrapyramidal, metabolic and sexual side effects being the most significant ones (Leucht et al. 2013). Several different preclinical models have been developed for these side effects. The catalepsy test is probably the most used model for assessing antipsychotic-induced extrapyramidal side effects (Arnt 1983). Many different versions of this test have been developed, but in general, they involve placing an animal in an uncomfortable position (such as with its forelimbs on a bar raised several cm from the floor) and the time it takes to step down from this position is recorded. While normal, untreated animals will have a very short latency (typically in the order of less than 3 s), virtually all antipsychotics induce a dose-dependent increase in step-down latency (Ellenbroek 1993). One of the few exceptions is clozapine (Arnt 1983), which at least in most cases does not induce overt catalepsy. This is interesting as clozapine is highly sedative, suggesting that the catalepsy model is able to distinguish between extrapyramidal motor effects and sedation. Another model for assessing extrapyramidal side effects was mentioned above, the paw test, where the forelimb retraction time is considered to be a model for this.

With respect to the metabolic side effects of antipsychotics, weight gain is one of the most troublesome effects as it occurs commonly with many antipsychotics (Goudie et al. 2005) and is particularly prevalent with clozapine and olanzapine (with body weight increases of more than 4 kg in 10 weeks time). Unfortunately, the pharmacology of antipsychotic weight gain is complex and involves many neurotransmitter systems and receptors including the histamine H1, the 5-HT2C and the 5-HT1A receptors (Goudie et al. 2005). Like in humans, antipsychotics increase food intake in rats and alter lipid and glucose metabolism (Baptista et al. 2002b; Hartfield et al. 2003; Zarate et al. 2004). However, there is strong evidence that in rats, the effects are sex-dependent. Thus, the effects of sulpiride (Baptista et al. 2002a), haloperidol and olanzapine (Pouzet et al. 2003) were only seen in female rats. This questions the validity of the food intake model in rats, as there is no evidence of a sex-bias effect in patients treated with antipsychotics, and additionally, haloperidol generally does not induce weight gain in patients (van der Zwaal et al. 2014). Nonetheless, a recent study using long-acting risperidone and olanzapine confirmed significant weight gain in female rats accompanied by an increase in several lipogenic genes (Ersland et al. 2015).

Sexual dysfunction is another serious side effect of antipsychotic drugs. Although often underestimated (and underreported by patients), its prevalence rate has been estimated to be about 54 % in males and 30 % in women (Just 2015). The most often reported sexual problems in men are problems with achieving and maintaining an erection as well as reduced orgasm intensity. In females, it mostly consists of reduced orgasm quality, reduced ability to achieve an orgasm and pain during orgasm. Dopamine receptor blockade is suggested to play a major role in these side effects, particularly the blockade of dopamine D2 receptors at the level of the magnocellular cells in the paraventricular nucleus of the hypothalamus, leading to a strong increase in prolactin (Just 2015). However, an additional influence of especially serotonin receptors has also been proposed as serotonin plays an important role in the regulation of sexual behaviour (Snoeren et al. 2010, 2014). Studies in animals have supported these findings showing, for example, that acute administration and chronic administration of haloperidol and risperidone reduce mounting behaviours intromissions and ejaculations in male rats (Drago et al. 1997). Although this study seemed to suggest that the effects of risperidone were only seen at low but not high doses, subsequent studies were unable to replicate this, finding significant decreases in mounting frequency with repeated risperidone injections (Zhang et al. 2011), although the data confirmed that haloperidol induced stronger effects on male sexual behaviour, leading to significantly reduced intromissions and ejaculations (and subsequently increasing intromission and ejaculation latency). Interestingly, in the same study, chronic (21 days) quetiapine did not affect sexual behaviour in male rats, which draw into question the validity of the animal model as there is no evidence that quetiapine induces less sexual side effects than other antipsychotic drugs in patients (La Torre et al. 2013). Much less research has been done on the preclinical effects of antipsychotic drugs on female sexual behaviour. This is likely due to the fact that sexual behaviour in rodents (as indeed in most mammalian species) is largely initiated and dominated by the male.

One of the most intriguing side effects is the recently reported misuse of antipsychotics, most prominently quetiapine. In a recent review, more than 2000 cases were identified from the data base of the American Association for Poison Control Centers (Klein-Schwartz et al. 2014). This abuse is quite perplexing as antipsychotics, including quetiapine, are dopamine antagonists (see above), while virtually all drugs of abuse are (in)direct dopamine agonist, increasing the release of dopamine, especially in the nucleus accumbens (Di Chiara and Imperato 1988). Indeed, studies in rats (Arnt 1995) and monkeys (Ellenbroek et al. 1996) have shown that quetiapine actually reverses the behavioural effects of psychostimulants such as amphetamine. Likewise, repeated administration of quetiapine induces depolarization inactivation of the dopaminergic cells especially in the ventral tegmental area, leading to a shutdown of these cells (Goldstein et al. 1989). It thus seems unlikely that quetiapine is, in itself, reinforcing like traditional drugs of abuse, such as cocaine or (meth)amphetamine. However, a study in rats showed that quetiapine was self-administered albeit to a very modest level (around 7 infusions over a 2 h period) and induced some degree of conditioned place preference (Cha et al. 2013). On the other hand, in contrast to cocaine, quetiapine by itself was not self-administered by macaque monkeys (Brutcher and Nader 2015). Nonetheless, in the same study, it was found that quetiapine in some cases enhanced the reinforcing properties of cocaine. Thus, while quetiapine did not affect the response to a maximal cocaine dose, when added to a submaximal dose, cocaine response rates were increased. Unfortunately, this effect was only seen in four out of the six monkeys and thus urgently needs replication. The question remains why quetiapine is abused in humans (and perhaps in rodents as well). As this is dealt with in the paper by Montebello (2016), we will not address this in any great detail. Suffice to say that several different mechanisms (including an anxiolytic effect and sedative/hypnotic of quetiapine) have been proposed.

In summary, in our paper, we have given an overview of the most important preclinical effects of antipsychotics drugs, covering both the therapeutic as well as the side effects. Although antipsychotics have been used now for over 50 years and there are clear differences between different antipsychotics, the neurobiology behind many of the effects of these drugs is still largely unknown. This is especially true for the side effects, and important improvements in the preclinical models are necessary in order to better understand these effects and ultimately develop drugs for the treatment of schizophrenia with a better safety profile.

References

Aggernaes B, Glenthoj BY, Ebdrup BH et al (2010) Sensorimotor gating and habituation in antipsychotic-naive, first-episode schizophrenia patients before and after 6 months' treatment with quetiapine. Int J Neuropsychopharmacol 13:1383–1395

Arnt J (1982) Pharmacological specificity of conditioned avoidance response inhibition in rats: inhibition by neuroleptics and correlation to dopamine receptor blockade. Acta Pharmacol Toxicol 51:321–329

Arnt J (1983) Neuroleptic inhibition of 6,7-ADTN-induced hyperactivity after injections into the nucleus accumbens. Specificity and comparison with other models. Eur J Pharmacol 90:47–55

Arnt J (1995) Differential effects of classical and newer antipsychotics on the hypermotility induced by two dose levels of d-amphetamine. Eur J Pharmacol 283:55–62

Arnt J (2000) Screening models for antipsychotic drugs. In: Ellenbroek BA, Cools AR (eds) Atypical antipsychotics, edn. Birkhäuser Verlag, Basel, pp 99–119

Baptista T, LaCruz A, Paez X et al (2002a) The antipsychotic drug sulpiride does not affect bodyweight in male rats. Is insulin resistance involved? Eur J Pharmacol 447:91–98

Baptista T, LaCruz A, De Mendoza D et al (2002b) Body weight gain after administration of antipsychotic drugs. Pharmacopsychiatry 35:36

Braff D, Stone C, Callaway E et al (1978) Prestimulus effects of human startle reflex in normals and schizophrenics. Psychophysiology 15:339–343

Braff DL (2015) The importance of endophenotypes in schizophrenia research. Schizophr Res 163:1–8

Braff DL, Geyer MA (1990) Sensorimotor gating and schizophrenia. Human and animal model studies. Arch Gen Psychiatry 47:181–188

Brutcher RE, Nader MA (2015) Effects of quetiapine treatment on cocaine self-administration and behavioral indices of sleep in adult rhesus monkeys. Psychopharmacology 232:411–420

Carlsson A, Lindqvist M (1963) Effects of chlorpromazine or haloperidol on formation of 3-methoxytyramin and normetanephrine in mouse brain. Acta Pharmacol Toxicol 20:140–144

Cha HJ, Lee HA, Ahn JI et al (2013) Dependence potential of quetiapine: behavioral pharmacology in rodents. Biomol Ther (Seoul) 21:307–312

De Fazio P, Gaetano R, Caroleo M et al (2015) Rare and very rare adverse effects of clozapine. Neuropsychiatric Dis Treat 11:1995–2003

Di Chiara G, Imperato A (1988) Drugs abused by humans preferentially increase synaptic dopamine concentrations in the mesolimbic system of freely moving rats. Proc Natl Acad Sci USA 85:5274–5278

Drago F, Contarino A, Marino R et al (1997) Effects of acute or chronic administration of risperidone on motor and sexual behavior of male rats. Pharmacol Res 35:17–25

During S, Glenthoj BY, Andersen GS et al (2014) Effects of dopamine d2/d3 blockade on human sensory and sensorimotor gating in initially antipsychotic-naive, first-episode schizophrenia patients. Neuropsychopharmacology 39:3000–3008
Ellenbroek B, Cools AR (1988) The Paw test: an animal model for neuroleptic drugs which fulfils the criteria for pharmacological isomorphism. Life Sci 42:1205–1213
Ellenbroek BA (1993) Treatment of schizophrenia: a clinical and preclinical evaluation of neuroleptic drugs. Pharmacol Ther 57:1–78
Ellenbroek BA (2012) Psychopharmacological treatment of schizophrenia: what do we have and what could we get? Neuropharmacology 62:1371–1380
Ellenbroek BA (2015) Schizophrenia: animal models. In: Stolerman IP, Price LH (eds) Encyclopedia of psychopharmacology, edn, vol 2. Springer, Heidelberg, pp 1501–1508
Ellenbroek BA, Ghiabi B (2015) Do histamine receptor 3 antagonists have a place in the therapy for schizophrenia? Curr Pharm Des 21:3760–3770
Ellenbroek BA, Cesura AM (2015) Antipsychotics and the dopamine—serotonin connection. In: Celanire S, Poli S (eds) Topics in medicinal chemistry, vol 13. Springer, Heidelberg, pp 1–50
Ellenbroek BA, Artz MT, Cools AR (1991) The involvement of dopamine D1 and D2 receptors in the effects of the classical neuroleptic haloperidol and the atypical neuroleptic clozapine. Eur J Pharmacol 196:103–108
Ellenbroek BA, Prinssen EP, Cools AR (1994) The role of serotonin receptor subtypes in the behavioural effects of neuroleptic drugs. A paw test study in rats. Eur J Neurosci 6:1–8
Ellenbroek BA, Lubbers LJ, Cools AR (1996) Activity of "seroquel" (ICI 204,636) in animal models for atypical properties of antipsychotics: a comparison with clozapine. Neuropsychopharmacology 15:406–416
Ellenbroek BA, Peeters BW, Honig WM et al (1987) The paw test: a behavioural paradigm for differentiating between classical and atypical neuroleptic drugs. Psychopharmacology Berl 93:343–348
Ersland KM, Skrede S, Rost TH et al (2015) Antipsychotic-induced metabolic effects in the female rat: direct comparison between long-acting injections of risperidone and olanzapine. J Psychopharmacol 29:1280–1289
Farde L, Wiesel FA, Halldin C et al (1988) Central D2-dopamine receptor occupancy in schizophrenic patients treated with antipsychotic drugs. Arch Gen Psychiatry 45:71–76
Fusar-Poli P, Papanastasiou E, Stahl D et al (2015) Treatments of negative symptoms in schizophrenia: meta-analysis of 168 randomized placebo-controlled trials. Schizophr Bull 41:892–899
Geyer MA, Krebs-Thomson K, Braff DL et al (2001) Pharmacological studies of prepulse inhibition models of sensorimotor gating deficits in schizophrenic patients: a decade in review. Psychopharmacology 156:117–154
Goldstein JM, Litwin LC, Sutton EB et al (1989) Effects of ICI 169,369, a selective serotonin2 antagonist, in electrophysiological tests predictive of antipsychotic activity. J Pharmacol Exp Ther 249:673–680
Goudie AJ, Cooper GD, Halford JCG (2005) Antipsychotic-induced weight gain. Diabetes Obes Metab 7:478–487
Graham F (1975) The more or less startling effects of weak prepulses. Psychophysiology 12:238–248
Haase HJ (1954) About prevalence and interpretation of the psychomotor parkinson syndrome in megaphen or largactil treatment duration. Nervenarzt 25:486–492
Haase HJ, Floru L, Knaack M (1974) Clinical importance of neuroleptic threshold and its fine motor determination by handwriting test. J Int Med Res 2:321–330
Hartfield AW, Moore NA, Clifton PG (2003) Effects of clozapine, olanzapine and haloperidol on the microstructure of ingestive behaviour in the rat. Psychopharmacology 167:115–122
Ishibashi T, Horisawa T, Tokuda K et al (2010) Pharmacological profile of lurasidone, a novel antipsychotic agent with potent 5-hydroxytryptamine 7 (5-HT7) and 5-HT1A receptor activity. J Pharmacol Exp Ther 334:171–181

Janssen PAJ, Niemegeers CJE, Schellekens KHL (1965) Is it possible to predict the clinical effects of neuroleptic drugs (major tranquilizers) from animal data? Part I: neuroleptic activity spectra for rats. Arzneimittelforschung 15:104–117

Janssen PAJ, Niemegeers CJE, Schellekens KHL (1966) Is it possible to predict the clinical effects of neuroleptic drugs (major tranquilizers) from animal data? Part III: The subcutaneous and oral activity in rats and dogs of 56 neuroleptic drugs in the jumping box test. Arzneimittelforschung 16:339–346

Janssen PAJ, Niemegeers CJE, Schellekens KHL et al (1967) Is it possible to predict the clinical effects of neuroleptic drugs (major tranquilizers) from animal data? Part IV: An improved experimental design for measuring the inhibitory effects of neuroleptic drugs on amphetamine-induced "chewing" and "agitation" in rats. Arzneimittelforschung 17:841–854

Just MJ (2015) The influence of atypical antipsychotic drugs on sexual function. Neuropsychiatric Dis Treat 11:1655–1661

Kahn RS, Fleischhacker WW, Boter H et al (2008) Effectiveness of antipsychotic drugs in first-episode schizophrenia and schizophreniform disorder: an open randomised clinical trial. Lancet 371:1085–1097

Kegeles LS, Abi-Dargham A, Frankle WG et al (2010) Increased synaptic dopamine function in associative regions of the striatum in schizophrenia. Arch Gen Psychiatry 67:231–239

Kelley BM, Porter JH (1997) The role of muscarinic cholinergic receptors in the discriminative stimulus properties of clozapine in rats. Pharmacol Biochem Behav 57:707–719

Klein-Schwartz W, Schwartz EK, Anderson BD (2014) Evaluation of quetiapine abuse and misuse reported to poison centers. J Addict Med 8:195–198

La Torre A, Conca A, Duffy D et al (2013) Sexual dysfunction related to psychotropic drugs: a critical review part II: antipsychotics. Pharmacopsychiatry 46:201–208

Leucht S, Cipriani A, Spineli L et al (2013) Comparative efficacy and tolerability of 15 antipsychotic drugs in schizophrenia: a multiple-treatments meta-analysis. Lancet 382:951–962

Leysen JE (2000) Receptor profile of antipsychotics. In: Ellenbroek BA, Cools AR (eds) Atypical antipsychotics. Birkhäuser Verlag, Basel, pp 57–81

Mackeprang T, Kristiansen KT, Glenthoj BY (2002) Effects of antipsychotics on prepulse inhibition of the startle response in drug-naive schizophrenic patients. Biol Psychiatry 52:863–873

Madras BK (2013) History of the discovery of the antipsychotic dopamine D2 receptor: a basis for the dopamine hypothesis of schizophrenia. J Hist Neurosci 22:62–78

Mansbach RS, Geyer MA, Braff DL (1988) Dopaminergic stimulation disrupts sensorimotor gating in the rat. Psychopharmacol Berl 94:507–514

Matthysse S (1986) Animal models in psychiatric research. Prog Brain Res 65:259–270

Meincke U, Morth D, Voss T et al (2004) Prepulse inhibition of the acoustically evoked startle reflex in patients with an acute schizophrenic psychosis—A longitudinal study. Eur Arch Psychiatry Clin Neurosci 254:415–421

Meltzer HY, Matsubara S, Lee JC (1989) The ratios of serotonin2 and dopamine2 affinities differentiate atypical and typical antipsychotic drugs. Psychopharmacol Bull 25:390–392

Moncrieff J (2013) Magic bullets for mental disorders: the emergence of the concept of an "antipsychotic" drug. J Hist Neurosci 22:30–46

Montebello ME, Brett J (2016) Misuse and associated harms of quetiapine and other atypical antipsychotics. Curr Topics Behav Neurosci. doi:10.1007/7854_2015_424

Nielsen EB (1988) Cholinergic mediation of the discriminative stimulus properties of clozapine. Psychopharmacology 94:115–118

Niemegeers CJ, Janssen PA (1979) A systematic study of the pharmacological activities of dopamine antagonists. Life Sci 24:2201–2216

Nord M, Farde L (2011) Antipsychotic occupancy of dopamine receptors in schizophrenia. CNS Neurosci Ther 17:97–103

Ogren SO, Hall H, Kohler C et al (1986) The selective dopamine D2 receptor antagonist raclopride discriminates between dopamine-mediated motor functions. Psychopharmacology 90:287–294

Ogren SO, Hall H, Kohler C et al (1984) Remoxipride, a new potential antipsychotic compound with selective antidopaminergic actions in the rat brain. Eur J Pharmacol 102:459–474
Porter JH, Prus AJ (2009) Discriminative stimulus properties of atypical and typical antipsychotic drugs: a review of preclinical studies. Psychopharmacology 203:279–294
Pouzet B, Mow T, Kreilgaard M et al (2003) Chronic treatment with antipsychotics in rats as a model for antipsychotic-induced weight gain in human. Pharmacol Biochem Behav 75:133–140
Quednow BB, Wagner M, Westheide J et al (2006) Sensorimotor gating and habituation of the startle response in schizophrenic patients randomly treated with amisulpride or olanzapine. Biol Psychiatry 59:536–545
Snoeren E, Chan J, Bovens A et al (2010) Serotonin transporter null mutation and sexual behavior in female rats: 5-HT1A receptor desensitization. J Sex Med 7:2424–2434
Snoeren EMS, Veening JG, Olivier B et al (2014) Serotonin 1A receptors and sexual behavior in female rats: a review. Pharmacol Biochem Behav 121:43–52
Snyder SH, Greenberg D, Yamamura J (1974) Anti-schizophrenic drugsand brain cholinergic receptors. Arch Gen Psychiatry 31:58–61
Steck H (1954) Le syndrome extra-pyramidal et di-encephalique au cours des traitments au Largactil et au Serpasil. Ann Med Psychol 112:737–743
Stewart J (1962) Differential responses based on the physiological consequences of pharmacological agents. Psychopharmacologia 3:132–138
Thomas KV, Handley SL (1978) Modulation of dexamphetamine-induced compulsive gnawing—including the possible involvement of presynaptic alpha-adrenoreceptors. Psychopharmacology 56:61–67
van der Zwaal EM, Janhunen SK, la Fleur SE et al (2014) Modelling olanzapine-induced weight gain in rats. Int J Neuropsychopharmacol 17:169–186
van Rossum J (1966) The significance of dopamine-receptor blockade for the mechanism of action of neuroleptic drugs. Arch Int Pharmacodyn Ther 160:492–494
Wadenberg MG, Hicks PB (1999) The conditioned avoidance response test re-evaluated: is it a sensitive test for the detection of potentially atypical antipsychotics? Neurosci Biobehav Rev 23:851–862
Zarate JM, Boksa P, Baptista T et al (2004) Effects of clozapine on behavioral and metabolic traits relevant for schizophrenia in two mouse strains. Psychopharmacology 171:162–172
Zhang XR, Zhang ZJ, Jenkins TA et al (2011) The dose-dependent effect of chronic administration of haloperidol, risperidone, and quetiapine on sexual behavior in the male rat. J Sex Med 8:3345–3353

Contribution of Impulsivity and Serotonin Receptor Neuroadaptations to the Development of an MDMA ('Ecstasy') Substance Use Disorder

Susan Schenk and Dane Aronsen

Abstract As is the case with other drugs of abuse, a proportion of ecstasy users develop symptoms consistent with a substance use disorder (SUD). In this paper, we propose that the pharmacology of MDMA, the primary psychoactive component of ecstasy tablets, changes markedly with repeated exposure and that neuroadaptations in dopamine and serotonin brain systems underlie the shift from MDMA use to MDMA misuse in susceptible subjects. Data from both the human and laboratory animal literature are synthesized to support the idea that (1) MDMA becomes a less efficacious serotonin releaser and a more efficacious dopamine releaser with the development of behaviour consistent with an SUD and (2) that upregulated serotonin receptor mechanisms contribute to the development of the MDMA SUD via dysregulated inhibitory control associated with the trait of impulsivity.

Keywords MDMA · Ecstasy · Substance use disorder · Serotonin · Dopamine · Inhibitory control · Impulsivity

Contents

Most drugs are initially used in a relatively controlled manner, but over time a pattern of uncontrolled and compulsive use can develop. This change in drug-taking from controlled to uncontrolled is accompanied by the development of pronounced drug-seeking behaviour. When someone takes a drug repeatedly, they form a strong association between the drug and its rewarding properties. The initial drug-taking is based primarily on these rewarding drug effects. With continued factors that accompany drug-taking (images, smells, people, paraphernalia) become associated

---

S. Schenk (✉) · D. Aronsen
School of Psychology, Victoria University of Wellington, Wellington, New Zealand
e-mail: Susan.Schenk@vuw.ac.nz

Curr Topics Behav Neurosci (2017) 34:17–32
DOI 10.1007/7854_2015_421
Published Online: 31 December 2015

with that same, rewarding, response. Mere exposure to these cues can then produce an intense craving and relapse, even in detoxified users. This is perhaps the most insidious aspect of drug misuse—the craving for drugs that leads to drug-taking and then more drug-seeking, leading to the vicious cycle of uncontrolled use that defines a substance use disorder (SUD).

We know that the shift to drug misuse is accompanied by changes in brain systems underlying motivation, learning, memory and reinforcement (Volkow and Baler 2014). Many studies have identified the brain chemical, dopamine (DA), as critical for the development and maintenance of drug-taking (Koob and Volkow 2010; Weiss et al. 1992), and activity in DA systems codes the significance of rewarding events. With repeated drug exposures, the contexts in which drugs are experienced begin to play a role in drug-seeking and drug-taking (Di Ciano and Everitt 2004; Everitt et al. 2008). Responses to physiologically relevant stimuli are decreased, while responses to stimuli that have been associated with drug-taking become enhanced (Everitt and Robbins 2005). Drugs of abuse begin to activate 'habit' circuitry in the brain that automatically, and preferentially, guides behaviour towards drugs (Everitt and Robbins 2005; Kalivas 2008). At the same time, there is a failure to inhibit behaviours that are associated with pathological drug-seeking (Kalivas 2008). Thus, there are persistent changes in brain circuitry so that drug-related stimuli initiate drug-seeking behaviour, while brain circuits that underlie inhibitory control and executive functioning become dysfunctional.

Acute exposure to MDMA increases synaptic levels of DA, like other amphetamines. It also stimulates the release of 5-HT, like other amphetamines. Unlike other amphetamines, however, MDMA preferentially stimulates release of serotonin (5-HT) with much smaller increases in synaptic DA (see Schenk 2011 for a review). Selective 5-HT uptake inhibitors are not abused (Lichtigfeld and Gillman 1998), and the reinforcing potency of amphetamine analogues was negatively correlated with affinity for the 5-HT transporter (Wee et al. 2005; Ritz and Kuhar 1989). Increasing synaptic 5-HT by addition of the releasing stimulant, fenfluramine, decreased amphetamine self-administration by primates (Wee and Woolverton 2006) supporting not only the idea that 5-HT releasing stimulants are not reinforcing but also that non-selective activation of 5-HT receptors is inhibitory to psychostimulant-produced reward (Rothman and Baumann 2006).

Evidence has accumulated that the 5-HT response does indeed limit the reinforcing effects of MDMA. Self-administration by rats proceeds more slowly than self-administration of other amphetamine-type stimulants, and a relatively large percentage of rats fail to acquire MDMA self-administration (Schenk et al. 2007). We wondered whether this reflected differences in the effect of MDMA on 5-HT and reasoned that rats that were most responsive to this neurochemical effect might be less likely to self-administer MDMA. This hypothesis was supported. MDMA-produced increases in synaptic 5-HT, as measured by in vivo microdialysis, predicted whether rats would learn to self-administer MDMA or not (Bradbury et al. 2014); rats that failed to acquire self-administration were more responsive to the MDMA-produced increase in 5-HT. Acquisition of self-administration by laboratory rats was facilitated by a neurotoxic, 5,7 DHT lesion (Bradbury et al. 2014) or by a mutation that prevented functioning of the 5-HT transporter (Oakly et al. 2014). These converging lines of evidence support the idea that 5-HT is inhibitory to MDMA self-administration.

The large increase in synaptic 5-HT, relative to DA, produced following administration of MDMA would, therefore, be expected to limit the development of an SUD. Several studies that applied DSM IV criteria, however, provided convincing evidence of MDMA dependence (Smith et al. 2014; Cottler et al. 2001; Degenhardt et al. 2010; Topp et al. 1999; Yen and Hsu 2007; McKetin et al. 2014). The DSM V no longer considers drug dependence but, instead, recognizes SUDs resulting from the use of 10 separate classes of drugs. MDMA is not amongst those drugs that are specifically acknowledged in the DSM, but use can, nonetheless, produce many of the symptoms that the DSM includes in the diagnosis of an SUD. Of the 11 symptoms that are used to define an SUD, at least 5 apply to some users of MDMA. For example, some users reported (1) taking the substance in larger amounts or for longer than they meant to (Soar et al. 2006; Parrott 2013), (2) cravings and urges to use the substance (Davis and Rosenberg 2014; Hopper et al. 2006; Huxster et al. 2006), (3) continuing to use, even when they knew they had a physical or psychological problem that could have been caused or made worse by the substance (Cottler et al. 2001, 2009; Yen and Hsu 2007), (4) tolerance to the positive effects (Yen and Hsu 2007; Parrott 2005; Kirkpatrick et al. 2014; Peroutka et al. 1988) and (5) development of withdrawal symptoms (McKetin et al. 2014; Peroutka et al. 1988; Curran and Travill 1997). The potential to develop an SUD is somewhat of a puzzle given the pharmacology of MDMA. Our thesis is that repeated exposure to MDMA renders the drug more comparable to stimulant drugs and that this change is required for the development of an MDMA SUD.

The human literature supports this idea. In experienced users, MDMA produced subjective effects that overlapped with those of both the 5-HT releasing stimulant, mCPP, and amphetamine (Tancer and Johanson 2001, 2003). The reinforcing effects, however, were only observed following administration of MDMA and amphetamine; mCPP did not produce positive effects, supporting the idea that stimulated 5-HT does not mediate the positive effects of MDMA. This idea was further reinforced by the demonstration that the dopamine antagonist, haloperidol, but not the 5-HT antagonist, ketanserin, blocked the positive mood produced following MDMA administration to healthy individuals with virtually no previous MDMA use as well as in experienced users (Liechti and Vollenweider 2000; Liechti et al. 2000). Some experienced users had difficulty differentiating the subjective effects of amphetamine and MDMA (Johanson et al. 2006) or methamphetamine and MDMA (Kirkpatrick et al. 2012), and MDMA administered to experienced users produced positive effects that were comparable to effects produced by other amphetamines (Kirkpatrick et al. 2012).

Because these studies in humans are almost always carried out in experienced users, there is limited information on whether the subjective effects change as a function of repeated drug use. This might be important because repeated exposure to various amphetamine-type drugs can markedly alter the pharmacodynamic responses, with some responses becoming sensitized and others becoming desensitized. Indeed, there is ample evidence of neuroplasticity derived from preclinical studies of the effects of repeated exposure to psychostimulant drugs, including MDMA. We propose that there are persistent alterations in the pharmacodynamic profile due to both 5-HT and DA neuroplasticity and that these changes underlie the development of an MDMA SUD.

An obvious prerequisite for the development of any SUD is drug exposure. How much drug exposure and the required conditions of the exposure are still unknown. It has become clear, however, that repeated intermittent exposure to psychostimulants often initiates a cascade of neurochemical events that renders some users susceptible to an SUD. The increase in synaptic DA, particularly in the nucleus accumbens shell, is a crucial initial response for coding the rewarding effect of a drug and enabling the development of associative learning that is critical for drug-seeking expressed in response to stimuli or contexts associated with the rewarding effect of a drug. Following repeated, intermittent exposure to most psychostimulant drugs, this DA response becomes sensitized (Robinson and Becker 1986) and this sensitized response might underlie the preferential attention to drugs of abuse over other, natural, reinforcers.

The nucleus accumbens also receives glutamatergic input from the prefrontal cortex and amygdala (Voorn et al. 2004). Thus, there are converging DA and glutamatergic inputs in the nucleus accumbens that modulates the response to psychostimulant drugs. Increased drug-produced DA in hippocampus and amygdala also strengthens the relationship between a response and the delivery of the drug reinforcer, i.e. learning stimulus–reward associations. Additionally, midbrain DA neurons project to cortical sites, and pyramidal neurons of the mPFC send glutamatergic projections to the ventral tegmental area, the nucleus accumbens and the amygdala (Gabbott et al. 2005; Hoover and Vertes 2007), further modulating those neurochemical responses. Increased DA in prefrontal, anterior cingulate and orbitofrontal regions impacts cognitive control and executive functioning via modulation of glutamatergic outputs to limbic regions.

Following repeated, intermittent exposure to many psychostimulant drugs, a number of neuroadaptations occur that are relevant to the preferential attendance to drugs and drug-related stimuli and the loss of inhibitory control regarding drug-seeking in response to those stimuli. The impact of cue presentation on drug-seeking might reflect, at least in part, a shift in the stimulus that elicits the DA response from the drug to anticipation of the drug effect (Schultz 1997). This suggests a strengthening of signals that mediate conditioned reinforcement and a corresponding decrease in the strength of the signal that mediates primary drug reinforcement at the level of the ventral striatum. At the same time, there is an inability to inhibit the behaviour that will result in drug-taking, suggesting a decrease in executive functioning and control at the level of the prefrontal, orbitofrontal and anterior cingulate cortices.

Some studies from our laboratory have confirmed the important role of the continued presentation of a stimulus associated with self-administered infusions in drug-taking behaviour. Rats were trained to self-administer cocaine (Schenk and Partridge 2001) or MDMA (Daniela et al. 2006), and each self-administered infusion was paired with the illumination of a stimulus light. Following extensive experience, drug infusions delivered contingent on an operant response, but without the associated light stimulus were relatively ineffective in maintaining operant responding. Importantly, the light stimulus acquired conditioned reinforcing properties that maintained responding in the absence of the drug reinforcer, suggesting a shift from drug-maintained responding to stimulus-maintained responding. Responding under a second-order schedule also demonstrates the high rate of responding that is maintained when only a drug-associated stimulus is presented (Schindler et al. 2002; Everitt and Robbins 2000).

Additionally, a large number of studies have documented the ability of drug-associated stimuli to elicit craving in abusers (Childress et al. 1993; Jasinska et al. 2014) and drug-seeking in some animal models (Crombag et al. 2008; Marchant et al. 2014; Grimm et al. 2001; Tzschentke 1998). Alterations in prefrontal cortical glutamatergic outputs to the nucleus accumbens (Kalivas 2009) and other downstream outputs that might include increased activation of dorsal striatal circuitry (Everitt and Robbins 2005, 2013) are posited to be of particular relevance to persistent drug-seeking behaviour in response to exposure to these drug-associated stimuli. These glutamatergic outputs are modulated by a large number of neurochemical systems including GABA, DA and 5-HT (Volkow and Baler 2014).

MDMA, like other psychostimulant drugs of abuse, increased DA preferentially in the nucleus accumbens shell (Cadoni et al. 2005). Both the human (Liechti and Vollenweider 2000) and animal (Brennan et al. 2009) literature support the idea that the MDMA-produced increase in DA is important for the reinforcing effects. Effects of repeated exposure on brain DA are, however, inconsistent. There was no change in DA transporter density in ecstasy users (Reneman et al. 2002; McCann et al. 2008), but one study suggested increased DOPA uptake in former ecstasy users (Tai et al. 2011). Ecstasy, polydrug, users were less sensitive to some effects of the D2 agonist, bromocriptine (Gerra et al. 2003), suggesting downregulated DA D2 receptor mechanisms. Results of preclinical studies are also equivocal. Some preclinical studies have failed to observe MDMA-produced changes in DA neurochemistry following repeated exposure (Shortall et al. 2013; Reveron et al. 2010; Fantegrossi et al. 2004; Shankaran and Gudelsky 1999; Baumann et al. 2008), and others have shown an increase (Shortall et al. 2013; Ludwig et al. 2008; Kalivas et al. 1998; Colussi-Mas et al. 2010).

In contrast to these neurochemical studies, a large number of behavioural studies have suggested that repeated MDMA exposure sensitized DA substrates. Acute exposure to MDMA produces hyperactivity that became sensitized following repeated exposure (Modi et al. 2006; Yang et al. 2011; Bradbury et al. 2012; Colussi-Mas and Schenk 2008; Schenk and Bradbury 2015; Ball et al. 2009, 2010, 2011) even in SERT knock out rats (Lizarraga et al. 2014). The locomotor-activating effects of MDMA have been shown to be dependent upon DA activation (Gold et al. 1989; Bubar et al. 2004; Daniela et al. 2004), although 5-HT mechanisms have also been demonstrated, especially in hyperactivity produced following administration of (+) MDMA (Callaway et al. 1990; Bankson and Cunningham 2002; McCreary et al. 1999). Additionally, some studies have shown cross-sensitization between the behavioural effects of MDMA and other psychostimulant drugs (Bradbury et al. 2012; Moreno-Sanz et al. 2009; Fletcher et al. 2001; Schenk et al. 2003). Importantly, drug-seeking following extensive MDMA self-administration was produced by DA, but not 5-HT agonists, and was attenuated by DA antagonists (Carati and Schenk 2011). Thus, although there is limited direct evidence of effects of repeated MDMA exposure on DA neurochemistry, the behavioural data are consistent with sensitized dopamine responses. A question remains as to whether this sensitized behavioural response reflects direct effects on DA neurotransmission or indirect effects resulting from other MDMA-produced neuroadaptations.

There is substantial evidence of alterations in 5-HT neurochemistry following repeated exposure to MDMA. Acute exposure to MDMA produces marked increases in synaptic 5-HT activating 5-HT receptors in a non-selective manner. Following repeated exposure to MDMA, there are many indices of compromised 5-HT neurotransmission. In particular, density of SERT is decreased in ecstasy users (Parrott 2013; McCann et al. 2008; Benningfield and Cowan 2013; Di Iorio et al. 2012; Parrott et al 2014; McCann et al. 2000) (but see (Cowan 2007) for a review of earlier human neuroimaging studies) and in rats and non-human primates (Schenk et al. 2007; Biezonski and Meyer 2011; Capela et al. 2009; Cole and Sumnall 2003; Ricaurte et al. 2000) that were exposed to MDMA. This decrease might produce additional alterations in receptor mechanisms that directly mediate the development of an SUD.

'Those who may have predicted that molecular biology would largely simplify the life of pharmacologists have missed the point for 5-HT research' (Hoyer et al. 2002). There are at least 14 different 5-HT receptor subtypes in 7 families of receptors (5-$HT_{1\text{-}7}$). Some of these receptors are localized on DA-containing nerve cell bodies and terminals or on cell bodies and terminals of neurons that modulate DA responses (Alex and Pehek 2007). For example, 5-HT_{1B}, 5-HT_{2A} and 5-HT_{2C} have been localized to the ventral tegmental area (Doherty and Pickel 2000), nucleus accumbens (Mengod et al. 1990) and prefrontal cortex (Mengod et al. 1990; Pazos and Palacios 1985). They are, therefore, well positioned to impact DA neurotransmission directly or to modify the impact of other brain systems on DA (Di Giovanni et al. 2010; Esposito et al. 2008). Some are also of interest because they are highly localized to prefrontal cortical substrates, activation alters glutamate signalling in the prefrontal cortex, and activation also impacts some measures of impulsivity/compulsivity (Homberg 2012; Clark et al. 2004). MDMA decreased glutamate-stimulated firing of nucleus accumbens cells (White et al. 1994; Obradovic et al. 1996). One study has demonstrated hypofrontality in MDMA abusers (Bosch et al. 2013) that might reflect the decreased SERT binding and ensuing altered 5-HT receptor mechanisms.

Reductions in SERT following repeated MDMA were accompanied by increased 5-HT_{2a} receptor binding (Benningfield and Cowan 2013; Urban et al. 2012). It appears likely that the increase in 5-HT_{2a} receptors is secondary to the decrease in SERT since chronic 5-HT depletion induced by tryptophan depletion in rats (Cahir et al. 2007) or neurotoxic 5,7 DHT lesions to mice (Heal et al. 1985) also increased 5-HT_{2a} receptor binding. This receptor is densely localized on PFC pyramidal neurons, and activation regulates glutamate release (Aznar and Klein 2013). It is possible that the effects of upregulated 5-HT2a receptor mechanisms on impulse control might relate to compulsive drug-seeking consistent with an MDMA SUD.

A number of preclinical models have been developed to investigate impulsive behaviour in animals (Pattij and Vanderschuren 2008), but the most widely used with respect to drug addiction are measures of behavioural inhibition, such as the 5-choice serial-reaction time task (5CSRTT), and measures of choice preference for a delayed reward, such as the delay-discounting paradigm. These measures show good validity as they are variants of those used to assess aspects of impulsive behaviour in humans (Robbins 2002; Evenden 1999).

Measures of behavioural inhibition model 'response impulsivity'. An animal is typically required to perform an operant in order to receive a reinforcer, but this operant must be performed after some delay. Operant responses that are made before the appropriate time are an index of response impulsivity. Most measures of response impulsivity employ a variation of a reaction time task. Reaction time tasks require an animal to maintain some response (e.g. hold down a lever) until a stimulus (light, tone) indicates that ceasing the response will produce a food reward (Blokland 1998). Impulsivity is operationalized as premature cessation of the required response. The 5CSRTT is the most widely used reaction time task and the most studied measure of response impulsivity in drug addiction. In this task, an animal is required to wait for a light stimulus to appear in one of five apertures before performing a nosepoke in that aperture. A nosepoke in an aperture before the light stimulus is presented is a premature response and indicates response impulsivity (Robbins 2002). The time between presentation of light stimuli (the intertrial interval, ITI) can be varied, with longer ITIs producing more premature responses.

Measures of choice preference for a delayed reward model 'decision impulsivity'. In the delay-discounting paradigm, an animal is required to choose between a small, immediate reinforcer and a larger, but delayed, reinforcer. At short delays, there is a preference for the large reinforcer, but as the delay for the larger reinforcer is increased, a preference for the immediate reinforcement develops. The delay following which there is no longer a preference for the larger, delayed reinforcer, indicated by a 50 % chance of choosing either the small or large reward, is defined as the 'indifference point'. Lower indifference points, showing intolerance to delay, are indicative of greater impulsivity. Indifference points for various delays can be obtained, and an indifference curve can be generated, by plotting the indifference point against delay, with a faster rate of decay reflecting greater impulsivity (Evenden 1999).

Delay discounting and reaction time tasks can be used to determine impulsivity scores across a group of animal subjects, which can then be divided into 'low impulsive' (LI) groups and 'high impulsive' (HI) groups. HI subjects are usually defined as those in the upper quartile of impulsivity scores, with LI subjects being those with impulsivity scores in the bottom quartile. These two groups can then be compared to see the impact of impulsivity on drug self-administration.

A role of 5-HT_{2a} receptors in both measures of impulsivity has been demonstrated. Systemic administration of the $5\text{-HT}_{2a/2c}$ agonist, 2,5-Dimethoxy-4-iodoamphetamine (DOI), increased premature responding on the 5CSRTT (Koskinen and Sirviö 2001; Koskinen et al. 2003), and this effect was blocked by the 5-HT_{2a} antagonist, ketanserin (Koskinen et al. 2000a, b), suggesting a 5-HT_{2a} effect. DOI administration also increased impulsive behaviour in the choice reaction time task (Blokland et al. 2005) and increased preference for the small, immediate reinforcer in the delay-discounting paradigm (Evenden and Ryan 1999), indicating greater impulsivity. Furthermore, blockade of 5-HT_{2a} receptors decreased impulsive responses. The 5-HT_{2a} antagonists, ketanserin (Passetti et al. 2003; Talpos et al. 2006; Fletcher et al. 2007; Ruotsalainen et al. 1997) and M100907, (Fletcher et al. 2007; Winstanley et al. 2004) decreased premature responses on the 5CSRTT, or the similar 1CSRTT (Anastasio et al. 2011). HI rats, defined by performance on the 1CSRTT, showed a greater behavioural response to DOI and showed greater 5-HT_{2a} receptor binding in the frontal cortex (Fink et al. 2015). Finally, premature responses were correlated with 5-HT_{2a} protein levels (Anastasio et al. 2015). These findings illustrate a likely role of 5-HT_{2a} receptor activation in impulsive behaviour, but are unfortunately limited by the lack of selective pharmacological tools, particularly agonists, targeting 5-HT_{2a} receptors. As more selective ligands are produced, the specific role of 5-HT_{2a} receptor activation on impulsive behaviour will be able to be further elucidated.

Based on these findings, an increase in 5-HT_{2a} receptors associated with MDMA exposure could increase impulsivity, thereby increasing compulsive drug-taking consistent with an SUD. The role of impulsivity in different aspects of drug self-administration in animals has been determined, but most research has focused on drugs other than MDMA. Some studies have looked at the acquisition and maintenance of self-administration, based on the idea that HI subjects might be more prone to take drugs (Perry and Carroll 2008). When impulsivity was determined using a delay discounting task, HI rats consumed more ethanol (Poulos et al. 1995), or cocaine (Perry et al. 2005, 2008; Koffarnus and Woods 2013) and cocaine self-administration were acquired more quickly and in a higher percentage of HI rats (Perry et al. 2005; Zlebnik and Carroll 2015). Impulsivity measured by performance in the 5CSRTT did not, however, predict acquisition of cocaine self-administration (Belin et al. 2008; Dalley et al. 2007). This difference might reflect different mechanisms for the two forms of impulsivity. On the other hand, rats that were HI as measured by 5CSRRT performance acquired nicotine self-administration more readily (Diergaarde et al. 2008), and a strain of mice with high impulsivity showed enhanced ethanol self-administration (Loos et al. 2013). Following acquisition, HI rats self-administered more cocaine per hour than LI rats and exhibited an upward shift in the cocaine dose–response curve (Dalley et al. 2007).

The reinstatement of drug-seeking after abstinence or extinction is an animal model of relapse that has also been associated with impulsivity (Perry and Carroll 2008; Cunningham and Anastasio 2014). Reinstatement of cocaine seeking was greater in HI rats when characterized by either the 5CSRTT (Economidou et al.

2009), or a delay discounting task (Perry et al. 2008), and one study found that HI rats met more DSM-IV-like criteria for cocaine dependence (Belin et al. 2008). HI rats, when defined by delay discounting, showed greater reinstatement of nicotine seeking than LI rats, and when impulsivity was defined by the 5CSRTT, HI rats showed greater resistance to extinction (Diergaarde et al. 2008). Furthermore, HI rats were less responsive than LI rats to pharmacological treatment to reduce cocaine seeking (Regier et al. 2014). Along with greater ethanol consumption and seeking, alcohol preferring ‘P’ rats show greater delay discounting (Beckwith and Czachowski 2014), while the highly impulsive BXD15 mice showed greater reinstatement of ethanol seeking (Loos et al. 2013).

To our knowledge, only one study has assessed the role of impulsivity on MDMA self-administration. Impulsivity as determined by the 5CSRTT did not predict the latency to acquire MDMA self-administration in rats, but it predicted the magnitude of the drug-seeking response for MDMA (Bird and Schenk 2013). In so far as drug-seeking is a correlate of propensity to relapse following abstinence, this finding suggests that impulsivity can be directly linked to an MDMA SUD.

In drug users, impulsivity is a risk factor for initiating drug-taking, for escalating drug use and for developing SUDs (Perry and Carroll 2008; De Wit 2009). For example, impulsive traits in youth and young adulthood positively predicted future drug use, an earlier onset of drug-taking, and the likelihood of developing an SUD (De Wit 2009; Sher et al. 2000; Tarter et al. 2007; Kirisci et al. 2007). The noted upregulation of 5-HT_{2a} receptors in MDMA users could be expected to contribute to impulsive decision-making. Indeed, there is some evidence that ecstasy users show heightened levels of impulsivity compared to non-ecstasy controls (Morgan 1998), but whether this abnormal level of 5-HT_{2a} receptors is a cause, or an effect, of MDMA use remains to be determined.

References

Alex KD, Pehek EA (2007) Pharmacologic mechanisms of serotonergic regulation of dopamine neurotransmission. Pharmacol Ther 113(2):296–320

Anastasio NC et al (2011) The serotonin (5-Ht) 5-Ht2a receptor: association with inherent and cocaine-evoked behavioral disinhibition in rats. Behav Pharmacol 22(3):248

Anastasio NC et al (2015) Serotonin (5-HT) 5-HT2A receptor (5-HT2AR): 5-HT2CR imbalance in medial prefrontal cortex associates with motor impulsivity. ACS Chem Neurosci 6(7): 1248–1258

Aznar S, Klein AB (2013) Regulating prefrontal cortex activation: an emerging role for the 5-HT (2)A serotonin receptor in the modulation of emotion-based actions? Mol Neurobiol 48 (3):841–853

Ball KT et al (2009) Sensitizing regimens of (+/−)3, 4-methylenedioxymethamphetamine (ecstasy) elicit enduring and differential structural alterations in the brain motive circuit of the rat. Neuroscience 160(2):264–274

Ball KT et al (2010) Electrophysiological and structural alterations in striatum associated with behavioral sensitization to (+/−)3,4-methylenedioxymethamphetamine (ecstasy) in rats: role of drug context. Neuroscience 171(3):794–811

Ball KT et al (2011) Behavioral sensitization to 3,4-methylenedioxymethamphetamine is long-lasting and modulated by the context of drug administration. Behav Pharmacol 22(8):847–850

Bankson MG, Cunningham KA (2002) Pharmacological studies of the acute effects of (+)-3,4-methylenedioxymethamphetamine on locomotor activity: role of 5-HT(1B/1D) and 5-HT (2) receptors. Neuropsychopharmacology 26(1):40–52

Baumann MH et al (2008) Tolerance to 3,4-methylenedioxymethamphetamine in rats exposed to single high-dose binges. Neuroscience 152(3):773–784

Beckwith S, Czachowski CL (2014) Increased delay discounting tracks with a high ethanol-seeking phenotype and subsequent ethanol seeking but not consumption. Alcohol Clin Exp Res 38(10):2607–2614

Belin D et al (2008) High impulsivity predicts the switch to compulsive cocaine-taking. Science 320(5881):1352–1355

Benningfield MM, Cowan RL (2013) Brain serotonin function in MDMA (ecstasy) users: evidence for persisting neurotoxicity. Neuropsychopharmacology 38(1):253–255

Biezonski DK, Meyer JS (2011) The nature of 3,4-methylenedioxymethamphetamine (MDMA)-induced serotonergic dysfunction: evidence for and against the neurodegeneration hypothesis. Curr Neuropharmacol 9(1):84–90

Bird J, Schenk S (2013) Contribution of impulsivity and novelty-seeking to the acquisition and maintenance of MDMA self-administration. Addict Biol 18(4):654–664

Blokland A (1998) Reaction time responding in rats. Neurosci Biobehav Rev 22(6):847–864

Blokland A, Sik A, Lieben C (2005) Evaluation of DOI, 8-OH-DPAT, eticlopride and amphetamine on impulsive responding in a reaction time task in rats. Behav Pharmacol 16 (2):93–100

Bosch OG et al (2013) Verbal memory deficits are correlated with prefrontal hypometabolism in (18)FDG PET of recreational MDMA users. PLoS ONE 8(4):e61234

Bradbury S, Gittings D, Schenk S (2012) Repeated exposure to MDMA and amphetamine: sensitization, cross-sensitization, and response to dopamine D(1)- and D(2)-like agonists. Psychopharmacology 223(4):389–399

Bradbury S et al (2014) Acquisition of MDMA self-administration: pharmacokinetic factors and MDMA-induced serotonin release. Addict Biol 19(5):874–884

Brennan KA et al (2009) Effect of D1-like and D2-like receptor antagonists on methamphetamine and 3,4-methylenedioxymethamphetamine self-administration in rats. Behav Pharmacol 20 (8):688–694

Bubar MJ et al (2004) Effects of dopamine D1- or D2-like receptor antagonists on the hypermotive and discriminative stimulus effects of (+)-MDMA. Psychopharmacology 173(3–4):326–336

Cadoni C et al (2005) Effect of 3,4-methylendioxymethamphetamine (MDMA, "ecstasy") on dopamine transmission in the nucleus accumbens shell and core. Brain Res 1055(1–2): 143–148

Cahir M et al (2007) Acute and chronic tryptophan depletion differentially regulate central 5-HT1A and 5-HT 2A receptor binding in the rat. Psychopharmacology 190(4):497–506

Callaway CW, Wing LL, Geyer MA (1990) Serotonin release contributes to the locomotor stimulant effects of 3,4-methylenedioxymethamphetamine in rats. J Pharmacol Exp Ther 254 (2):456–464

Capela JP et al (2009) Molecular and cellular mechanisms of ecstasy-induced neurotoxicity: an overview. Mol Neurobiol 39(3):210–271

Carati C, Schenk S (2011) Role of dopamine D1- and D2-like receptor mechanisms in drug-seeking following methamphetamine self-administration in rats. Pharmacol Biochem Behav 98(3):449–454

Childress AR et al (1993) Cue reactivity and cue reactivity interventions in drug dependence. NIDA Res Monogr 137:73–95

Clark L, Cools R, Robbins TW (2004) The neuropsychology of ventral prefrontal cortex: decision-making and reversal learning. Brain Cogn 55(1):41–53

Cole JC, Sumnall HR (2003) The pre-clinical behavioural pharmacology of 3,4-methylenedioxymethamphetamine (MDMA). Neurosci Biobehav Rev 27(3):199–217
Colussi-Mas J, Schenk S (2008) Acute and sensitized response to 3,4-methylenedioxymethamphetamine in rats: different behavioral profiles reflected in different patterns of Fos expression. Eur J Neurosci 28(9):1895–1910
Colussi-Mas J et al (2010) Drug-seeking in response to a priming injection of MDMA in rats: relationship to initial sensitivity to self-administered MDMA and dorsal striatal dopamine. Int J Neuropsychopharmacol 13(10):1315–1327
Cottler LB et al (2001) Ecstasy abuse and dependence among adolescents and young adults: applicability and reliability of DSM-IV criteria. Hum Psychopharmacol 16(8):599–606
Cottler LB, Leung KS, Abdallah AB (2009) Test-re-test reliability of DSM-IV adopted criteria for 3,4-methylenedioxymethamphetamine (MDMA) abuse and dependence: a cross-national study. Addiction 104(10):1679–1690
Cowan RL (2007) Neuroimaging research in human MDMA users: a review. Psychopharmacology 189(4):539–556
Crombag HS et al (2008) Review. Context-induced relapse to drug-seeking: a review. Philos Trans R Soc Lond B Biol Sci 363(1507):3233–3243
Cunningham KA, Anastasio NC (2014) Serotonin at the nexus of impulsivity and cue reactivity in cocaine addiction. Neuropharmacology 76:460–478
Curran HV, Travill RA (1997) Mood and cognitive effects of +/−3,4-methylenedioxymethamphetamine (MDMA, 'ecstasy'): week-end 'high' followed by mid-week low. Addiction 92(7):821–831
Dalley JW et al (2007) Nucleus accumbens D2/3 receptors predict trait impulsivity and cocaine reinforcement. Science 315(5816):1267–1270
Daniela E et al (2004) Effect of SCH 23390 on (+/−)-3,4-methylenedioxymethamphetamine hyperactivity and self-administration in rats. Pharmacol Biochem Behav 77(4):745–750
Daniela E, Gittings D, Schenk S (2006) Conditioning following repeated exposure to MDMA in rats: role in the maintenance of MDMA self-administration. Behav Neurosci 120(5): 1144–1150
Davis AK, Rosenberg H (2014) The prevalence, intensity, and assessment of craving for MDMA/ecstasy in recreational users. J Psychoactive Drugs 46(2):154–161
De Wit H (2009) Impulsivity as a determinant and consequence of drug use: a review of underlying processes. Addict Biol 14(1):22–31
Degenhardt L, Bruno R, Topp L (2010) Is ecstasy a drug of dependence? Drug Alcohol Depend 107(1):1–10
Di Ciano P, Everitt BJ (2004) Conditioned reinforcing properties of stimuli paired with self-administered cocaine, heroin or sucrose: implications for the persistence of addictive behaviour. Neuropharmacology 47(Suppl 1):202–213
Di Giovanni G, Esposito E, Di Matteo V (2010) Role of serotonin in central dopamine dysfunction. CNS Neurosci Ther 16(3):179–194
Di Iorio CR et al (2012) Evidence for chronically altered serotonin function in the cerebral cortex of female 3,4-methylenedioxymethamphetamine polydrug users. Arch Gen Psychiatry 69 (4):399–409
Diergaarde L et al (2008) Impulsive choice and impulsive action predict vulnerability to distinct stages of nicotine seeking in rats. Biol Psychiatry 63(3):301–308
Doherty MD, Pickel VM (2000) Ultrastructural localization of the serotonin 2A receptor in dopaminergic neurons in the ventral tegmental area. Brain Res 864(2):176–185
Economidou D et al (2009) High impulsivity predicts relapse to cocaine-seeking after punishment-induced abstinence. Biol Psychiatry 65(10):851–856
Esposito E, Di Matteo V, Di Giovanni G (2008) Serotonin-dopamine interaction: an overview. Prog Brain Res 172:3–6
Evenden JL (1999) Varieties of impulsivity. Psychopharmacology 146(4):348–361

Evenden J, Ryan C (1999) The pharmacology of impulsive behaviour in rats VI: the effects of ethanol and selective serotonergic drugs on response choice with varying delays of reinforcement. Psychopharmacology 146(4):413–421

Everitt BJ, Robbins TW (2000) Second-order schedules of drug reinforcement in rats and monkeys: measurement of reinforcing efficacy and drug-seeking behaviour. Psychopharmacology 153(1):17–30

Everitt BJ, Robbins TW (2005) Neural systems of reinforcement for drug addiction: from actions to habits to compulsion. Nat Neurosci 8(11):1481–1489

Everitt BJ, Robbins TW (2013) From the ventral to the dorsal striatum: devolving views of their roles in drug addiction. Neurosci Biobehav Rev 37(9):1946–1954

Everitt BJ et al (2008) Review. Neural mechanisms underlying the vulnerability to develop compulsive drug-seeking habits and addiction. Philos Trans R Soc Lond B Biol Sci 363 (1507):3125–3135

Fantegrossi WE et al (2004) Behavioral and neurochemical consequences of long-term intravenous self-administration of MDMA and its enantiomers by rhesus monkeys. Neuropsychopharmacology 29(7):1270–1281

Fink LH et al (2015) Individual differences in impulsive action reflect variation in the cortical serotonin 5-HT2A receptor system. Neuropsychopharmacology

Fletcher PJ, Robinson SR, Slippoy DL (2001) Pre-exposure to (+/−) 3,4-methylenedioxy-methamphetamine (MDMA) facilitates acquisition of intravenous cocaine self-administration in rats. Neuropsychopharmacology 25(2):195–203

Fletcher PJ et al (2007) Opposing effects of 5-HT2A and 5-HT2C receptor antagonists in the rat and mouse on premature responding in the five-choice serial reaction time test. Psychopharmacology 195(2):223–234

Gabbott PL et al (2005) Prefrontal cortex in the rat: projections to subcortical autonomic, motor, and limbic centers. J Comp Neurol 492(2):145–177

Gerra G et al (2003) Hypothalamic-pituitary-adrenal axis responses to stress in subjects with 3,4-methylenedioxy-methamphetamine ('ecstasy') use history: correlation with dopamine receptor sensitivity. Psychiatry Res 120(2):115–124

Gold LH, Hubner CB, Koob GF (1989) A role for the mesolimbic dopamine system in the psychostimulant actions of MDMA. Psychopharmacology 99(1):40–47

Grimm JW et al (2001) Neuroadaptation. Incubation of cocaine craving after withdrawal. Nature 412(6843):141–142

Heal DJ et al (1985) Intracerebroventricular administration of 5,7-dihydroxytryptamine to mice increases both head-twitch response and the number of cortical 5-HT2 receptors. Neuropharmacology 24(12):1201–1205

Homberg JR (2012) Serotonin and decision making processes. Neurosci Biobehav Rev 36(1): 218–236

Hoover WB, Vertes RP (2007) Anatomical analysis of afferent projections to the medial prefrontal cortex in the rat. Brain Struct Funct 212(2):149–179

Hopper JW et al (2006) Incidence and patterns of polydrug use and craving for ecstasy in regular ecstasy users: an ecological momentary assessment study. Drug Alcohol Depend 85(3): 221–235

Hoyer D, Hannon JP, Martin GR (2002) Molecular, pharmacological and functional diversity of 5-HT receptors. Pharmacol Biochem Behav 71(4):533–554

Huxster JK, Pirona A, Morgan MJ (2006) The sub-acute effects of recreational ecstasy (MDMA) use: a controlled study in humans. J Psychopharmacol 20(2):281–290

Jasinska AJ et al (2014) Factors modulating neural reactivity to drug cues in addiction: a survey of human neuroimaging studies. Neurosci Biobehav Rev 38:1–16

Johanson CE et al (2006) Discriminative stimulus effects of 3,4-methylenedioxymethamphetamine (MDMA) in humans trained to discriminate among d-amphetamine, meta-chlorophenylpiperazine and placebo. Drug Alcohol Depend 81(1):27–36

Kalivas PW (2008) Addiction as a pathology in prefrontal cortical regulation of corticostriatal habit circuitry. Neurotox Res 14(2–3):185–189

Kalivas PW (2009) The glutamate homeostasis hypothesis of addiction. Nat Rev Neurosci 10 (8):561–572

Kalivas PW, Duffy P, White SR (1998) MDMA elicits behavioral and neurochemical sensitization in rats. Neuropsychopharmacology 18(6):469–479

Kirisci L et al (2007) Developmental trajectory classes in substance use disorder etiology. Psychol Addict Behav 21(3):287

Kirkpatrick MG et al (2012) A direct comparison of the behavioral and physiological effects of methamphetamine and 3,4-methylenedioxymethamphetamine (MDMA) in humans. Psychopharmacology 219(1):109–122

Kirkpatrick MG et al (2014) MDMA effects consistent across laboratories. Psychopharmacology 231(19):3899–3905

Koffarnus MN, Woods JH (2013) Individual differences in discount rate are associated with demand for self-administered cocaine, but not sucrose. Addict Biol 18(1):8–18

Koob GF, Volkow ND (2010) Neurocircuitry of addiction. Neuropsychopharmacology 35(1): 217–238

Koskinen T, Sirviö J (2001) Studies on the involvement of the dopaminergic system in the 5-HT 2 agonist (DOI)-induced premature responding in a five-choice serial reaction time task. Brain Res Bull 54(1):65–75

Koskinen T, Ruotsalainen S, Sirviö J (2000a) The 5-HT 2 receptor activation enhances impulsive responding without increasing motor activity in rats. Pharmacol Biochem Behav 66(4): 729–738

Koskinen T et al (2000b) Activation of 5-HT 2A receptors impairs response control of rats in a five-choice serial reaction time task. Neuropharmacology 39(3):471–481

Koskinen T, Haapalinna A, Sirvi J (2003) α-adrenoceptor-mediated modulation of 5-HT2 receptor agonist induced impulsive responding in a 5-choice serial reaction time task. Pharmacol Toxicol 92(5):214–225

Lichtigfeld FJ, Gillman MA (1998) Antidepressants are not drugs of abuse or dependence. Postgrad Med J 74(875):529–532

Liechti ME, Vollenweider FX (2000) Acute psychological and physiological effects of MDMA ("Ecstasy") after haloperidol pretreatment in healthy humans. Eur Neuropsychopharmacol 10 (4):289–295

Liechti ME et al (2000) Psychological and physiological effects of MDMA ("Ecstasy") after pretreatment with the 5-HT(2) antagonist ketanserin in healthy humans. Neuropsychopharmacology 23(4):396–404

Lizarraga LE et al (2014) Serotonin reuptake transporter deficiency modulates the acute thermoregulatory and locomotor activity response to 3,4-(+/−)-methylenedioxymethamphetamine, and attenuates depletions in serotonin levels in SERT-KO rats. Toxicol Sci 139 (2):421–431

Loos M et al (2013) Enhanced alcohol self-administration and reinstatement in a highly impulsive, inattentive recombinant inbred mouse strain. Front Behav Neurosci 7

Ludwig V, Mihov Y, Schwarting RK (2008) Behavioral and neurochemical consequences of multiple MDMA administrations in the rat: role of individual differences in anxiety-related behavior. Behav Brain Res 189(1):52–64

Marchant NJ et al (2014) Role of corticostriatal circuits in context-induced reinstatement of drug-seeking. Brain Res

McCann UD, Eligulashvili V, Ricaurte GA (2000) (+/−)3,4-Methylenedioxymethamphetamine ('Ecstasy')-induced serotonin neurotoxicity: clinical studies. Neuropsychobiology 42(1):11–16

McCann UD et al (2008) Positron emission tomographic studies of brain dopamine and serotonin transporters in abstinent (+/−)3,4-methylenedioxymethamphetamine ("ecstasy") users: relationship to cognitive performance. Psychopharmacology 200(3):439–450

McCreary AC, Bankson MG, Cunningham KA (1999) Pharmacological studies of the acute and chronic effects of (+)-3,4-methylenedioxymethamphetamine on locomotor activity: role of 5-hydroxytryptamine(1A) and 5-hydroxytryptamine(1B/1D) receptors. J Pharmacol Exp Ther 290(3):965–973

McKetin R et al (2014) The effect of the ecstasy 'come-down' on the diagnosis of ecstasy dependence. Drug Alcohol Depend 139:26–32
Mengod G et al (1990) Localization of the mRNA for the 5-HT2 receptor by in situ hybridization histochemistry. Correlation with the distribution of receptor sites. Brain Res 524(1):139–143
Modi GM et al (2006) Chronic exposure to MDMA (Ecstasy) elicits behavioral sensitization in rats but fails to induce cross-sensitization to other psychostimulants. Behav Brain Funct 2:1
Moreno-Sanz G et al (2009) Administration of MDMA to ethanol-deprived rats increases ethanol operant self-administration and dopamine release during reinstatement. Int J Neuropsychopharmacol 12(7):929–940
Morgan MJ (1998) Recreational use of "ecstasy"(MDMA) is associated with elevated impulsivity. Neuropsychopharmacology 19(4):252–264
Oakly AC et al (2014) A genetic deletion of the serotonin transporter greatly enhances the reinforcing properties of MDMA in rats. Mol Psychiatry 19(5):534–535
Obradovic T, Imel KM, White SR (1996) Methylenedioxymethamphetamine-induced inhibition of neuronal firing in the nucleus accumbens is mediated by both serotonin and dopamine. Neuroscience 74(2):469–481
Parrott AC (2005) Chronic tolerance to recreational MDMA (3,4-methylenedioxymethamphetamine) or ecstasy. J Psychopharmacol 19(1):71–83
Parrott AC (2013) Human psychobiology of MDMA or 'Ecstasy': an overview of 25 years of empirical research. Hum Psychopharmacol 28(4):289–307
Parrott AC, MDMA is certainly damaging after 25 years of empirical research: A reply and refutation of Doblin et al (2014). Human Psychopharmacology, 2014. 29(2): p. 109–119
Passetti F, Dalley JW, Robbins TW (2003) Double dissociation of serotonergic and dopaminergic mechanisms on attentional performance using a rodent five-choice reaction time task. Psychopharmacology 165(2):136–145
Pattij T, Vanderschuren LJ (2008) The neuropharmacology of impulsive behaviour. Trends Pharmacol Sci 29(4):192–199
Pazos A, Palacios JM (1985) Quantitative autoradiographic mapping of serotonin receptors in the rat brain. I. Serotonin-1 receptors. Brain Res 346(2):205–230
Peroutka SJ, Newman H, Harris H (1988) Subjective effects of 3,4-methylenedioxymethamphetamine in recreational users. Neuropsychopharmacology 1(4):273–277
Perry JL, Carroll ME (2008) The role of impulsive behavior in drug abuse. Psychopharmacology 200(1):1–26
Perry JL et al (2005) Impulsivity (delay discounting) as a predictor of acquisition of IV cocaine self-administration in female rats. Psychopharmacology 178(2–3):193–201
Perry JL, Nelson SE, Carroll ME (2008a) Impulsive choice as a predictor of acquisition of IV cocaine self-administration and reinstatement of cocaine-seeking behavior in male and female rats. Exp Clin Psychopharmacol 16(2):165–177
Perry JL, Nelson SE, Carroll ME (2008b) Impulsive choice as a predictor of acquisition of IV cocaine self-administration and reinstatement of cocaine-seeking behavior in male and female rats. Experiment Clin Psychopharmacol 16(2):165
Poulos CX, Le A, Parker J (1995) Impulsivity predicts individual susceptibility to high levels of alcohol self-administration. Behav Pharmacol
Regier PS et al (2014) Cocaine-, caffeine-, and stress-evoked cocaine reinstatement in high vs. low impulsive rats: treatment with allopregnanolone. Drug Alcohol Depend 143:58–64
Reneman L et al (2002) Use of amphetamine by recreational users of ecstasy (MDMA) is associated with reduced striatal dopamine transporter densities: a [123I]beta-CIT SPECT study–preliminary report. Psychopharmacology 159(3):335–340
Reveron ME, Maier EY, Duvauchelle CL (2010) Behavioral, thermal and neurochemical effects of acute and chronic 3,4-methylenedioxymethamphetamine ("Ecstasy") self-administration. Behav Brain Res 207(2):500–507
Ricaurte GA, Yuan J, McCann UD (2000) (+/−)3,4-methylenedioxymethamphetamine ('Ecstasy')-induced serotonin neurotoxicity: studies in animals. Neuropsychobiology 42(1):5–10

Ritz MC, Kuhar MJ (1989) Relationship between self-administration of amphetamine and monoamine receptors in brain: comparison with cocaine. J Pharmacol Exp Ther 248(3): 1010–1017
Robbins T (2002) The 5-choice serial reaction time task: behavioural pharmacology and functional neurochemistry. Psychopharmacology 163(3–4):362–380
Robinson TE, Becker JB (1986) Enduring changes in brain and behavior produced by chronic amphetamine administration: a review and evaluation of animal models of amphetamine psychosis. Brain Res 396(2):157–198
Rothman RB, Baumann MH (2006) Balance between dopamine and serotonin release modulates behavioral effects of amphetamine-type drugs. Ann NY Acad Sci 1074:245–260
Ruotsalainen S et al (1997) Differential effects of three 5-HT receptor antagonists on the performance of rats in attentional and working memory tasks. Eur Neuropsychopharmacol 7 (2):99–108
Schenk S (2011) MDMA ("ecstasy") abuse as an example of dopamine neuroplasticity. Neurosci Biobehav Rev 35(5):1203–1218
Schenk S, Bradbury S (2015) Persistent sensitisation to the locomotor activating effects of MDMA following MDMA self-administration in rats. Pharmacol Biochem Behav 132:103–107
Schenk S, Partridge B (2001) Influence of a conditioned light stimulus on cocaine self-administration in rats. Psychopharmacology 154(4):390–396
Schenk S et al (2003) Development, maintenance and temporal pattern of self-administration maintained by ecstasy (MDMA) in rats. Psychopharmacology 169(1):21–27
Schenk S et al (2007) MDMA self-administration in rats: acquisition, progressive ratio responding and serotonin transporter binding. Eur J Neurosci 26(11):3229–3236
Schindler CW, Panlilio LV, Goldberg SR (2002) Second-order schedules of drug self-administration in animals. Psychopharmacology 163(3–4):327–344
Schultz W (1997) Dopamine neurons and their role in reward mechanisms. Curr Opin Neurobiol 7 (2):191–197
Shankaran M, Gudelsky GA (1999) A neurotoxic regimen of MDMA suppresses behavioral, thermal and neurochemical responses to subsequent MDMA administration. Psychopharmacology 147(1):66–72
Sher KJ, Bartholow BD, Wood MD (2000) Personality and substance use disorders: a prospective study. J Consult Clin Psychol 68(5):818
Shortall SE et al (2013) Behavioural and neurochemical comparison of chronic intermittent cathinone, mephedrone and MDMA administration to the rat. Eur Neuropsychopharmacol 23 (9):1085–1095
Smith JL et al (2014) Deficits in behavioural inhibition in substance abuse and addiction: a meta-analysis. Drug Alcohol Depend 145:1–33
Soar K, Turner JJ, Parrott AC (2006) Problematic versus non-problematic ecstasy/MDMA use: the influence of drug usage patterns and pre-existing psychiatric factors. J Psychopharmacol 20 (3):417–424
Tai YF et al (2011) Persistent nigrostriatal dopaminergic abnormalities in ex-users of MDMA ('Ecstasy'): an 18F-dopa PET study. Neuropsychopharmacology 36(4):735–743
Talpos JC, Wilkinson LS, Robbins TW (2006) A comparison of multiple 5-HT receptors in two tasks measuring impulsivity. J Psychopharm 20(1):47–58
Tancer ME, Johanson CE (2001) The subjective effects of MDMA and mCPP in moderate MDMA users. Drug Alcohol Depend 65(1):97–101
Tancer M, Johanson CE (2003) Reinforcing, subjective, and physiological effects of MDMA in humans: a comparison with d-amphetamine and mCPP. Drug Alcohol Depend 72(1):33–44
Tarter RE et al (2007) Modeling the pathways linking childhood hyperactivity and substance use disorder in young adulthood. Psychol Addict Behav 21(2):266
Topp L et al (1999) Ecstasy use in Australia: patterns of use and associated harm. Drug Alcohol Depend 55(1–2):105–115

Tzschentke TM (1998) Measuring reward with the conditioned place preference paradigm: a comprehensive review of drug effects, recent progress and new issues. Prog Neurobiol 56 (6):613–672

Urban NB et al (2012) Sustained recreational use of ecstasy is associated with altered pre and postsynaptic markers of serotonin transmission in neocortical areas: a PET study with [(1)(1)C] DASB and [(1)(1)C]MDL 100907. Neuropsychopharmacology 37(6):1465–1473

Volkow ND, Baler RD (2014) Addiction science: uncovering neurobiological complexity. Neuropharmacology 76:235–49

Voorn P et al (2004) Putting a spin on the dorsal-ventral divide of the striatum. Trends Neurosci 27 (8):468–474

Wee S, Woolverton WL (2006) Self-administration of mixtures of fenfluramine and amphetamine by rhesus monkeys. Pharmacol Biochem Behav 84(2):337–343

Wee S et al (2005) Relationship between the serotonergic activity and reinforcing effects of a series of amphetamine analogs. J Pharmacol Exp Ther 313(2):848–854

Weiss F et al (1992) Neurochemical correlates of cocaine and ethanol self-administration. Ann NY Acad Sci 654:220–241

White SR, Duffy P, Kalivas PW (1994) Methylenedioxymethamphetamine depresses glutamate-evoked neuronal firing and increases extracellular levels of dopamine and serotonin in the nucleus accumbens in vivo. Neuroscience 62(1):41–50

Winstanley CA et al (2004) 5-HT2A and 5-HT2C receptor antagonists have opposing effects on a measure of impulsivity: interactions with global 5-HT depletion. Psychopharmacology 176(3–4):376–385

Yang PB, Atkins KD, Dafny N (2011) Behavioral sensitization and cross-sensitization between methylphenidate amphetamine, and 3,4-methylenedioxymethamphetamine (MDMA) in female SD rats. Eur J Pharmacol 661(1–3):72–85

Yen CF, Hsu SY (2007) Symptoms of ecstasy dependence and correlation with psychopathology in Taiwanese adolescents. J Nerv Ment Dis 195(10):866–869

Zlebnik NE, Carroll ME (2015) Effects of the combination of wheel running and atomoxetine on cue-and cocaine-primed reinstatement in rats selected for high or low impulsivity. Psychopharmacology 232(6):1049–1059

The Abuse Potential of Prescription Opioids in Humans—Closing in on the First Century of Research

Sharon L. Walsh and Shanna Babalonis

Abstract While opioids are very effective analgesics for treating acute pain, humans have struggled with opiate addiction for millenia. An opium abuse epidemic in the early 1900's led the US government to develop a systematic research infrastructure and scientific plan to produce new compounds with analgesic properties but without abuse liability. This review describes the techniques that were developed for testing in the human laboratory, including empirically derived outcome measures and required elements for human abuse potential assessment. The evaluation and characterization of semi-synthetic and synthetic opioids, including full mu opioid agonists, partial agonists and mixed agonist-antagonists, are described across several decades of research. Finally, the prescription opioid epidemic beginning in the 1990's in the US led to a resurgence in abuse potential evaluations, and the application of these methods to the study of novel abuse-deterrent formulations is discussed.

Keywords Opioid abuse · Abuse potential testing · Opioid agonist · Mixed agonist-antagonist · Atypical opioids · Abuse deterrence

Contents

S.L. Walsh (✉) · S. Babalonis
Department of Behavioral Science, Center on Drug and Alcohol Research, University of Kentucky, Lexington, USA
e-mail: sharon.walsh@uky.edu

Curr Topics Behav Neurosci (2017) 34:33–58
DOI 10.1007/7854_2016_448
Published Online: 30 June 2016

1 Brief Historical View of Opioid Abuse Potential Evaluation

Medicinal and non-medicinal use of opiates dates back millennia, and the agony of opiate addiction has been documented in numerous historical accountings (see, as one illustrative example, de Quincy 1823). It has been long recognized that opiates (the naturally occurring alkaloids from the opium plant) and opioids (inclusive of opiates as well as synthetic or semi-synthetic compounds), particularly those exerting their effects through the *mu* opioid receptors, are highly effective as analgesics especially for the treatment of acute pain. Unfortunately, their clinical use can be problematic because of the robust euphoric effects they can produce and the physiological dependence syndrome that ensues after repeated exposure that leads to great suffering upon drug withdrawal and serves to perpetuate continued use. It is this conundrum, the exquisite analgesic efficacy coupled with the abuse and dependence potential of opioids, that has spurred substantial research over the past century in search of the "holy grail" of pharmacology, the ideal opioid molecule that separates these intrinsic actions.

The advent of abuse potential testing may be traced to the 1920s, nearly 100 years ago now, when the United States (USA) Bureau of Social Hygiene and the National Research Council together established the Committee on Drug Addiction in order to respond to the growing opium addiction problem in the USA. It was anticipated that applying a carefully designed scientific approach could both eliminate the opium addiction problem and improve upon the available opiate analgesics by identifying novel opioids that would retain the desired analgesic efficacy while eliminating the abuse and dependence potential associated with the natural opiate agonists. Thus, the committee was tasked in 1929 with the "replacement of all present use of addiction alkaloids by substitutes having no addiction properties" (Eddy 1973). Three parallel scientific enterprises were initiated: (1) medicinal chemistry studies to create new molecules, (2) preclinical testing procedures to evaluate these new entities in vivo, and (3) clinical pharmacology studies of these compounds in human subjects. At that time, it was determined that the human testing programs would be conducted within the US Bureau of Prisons with prisoners—many of whom were incarcerated for substance abuse and/or related crimes. While this type of research in prisoners would be considered unethical in contemporary society, it was, at that time, a seemingly rational, acceptable, and convenient approach. Thus, a program was initiated through the US Public Health Service to conduct systematic studies of opioids and other drugs associated with abuse at the federal prison that came to be known as the Lexington Prison Farm (also referred to as the "Narcotics Farm" or "Narco"). The history of this place and the role it played in the field of clinical pharmacology in drug abuse are well documented in an excellent book for the interested reader (Campbell et al. 2008). There is a substantial body of interesting and important research that was conducted during these years; unfortunately, not all of it was published in the peer-reviewed literature as much of the early work was submitted in government reports that are available now only as archived works. It is worthwhile noting that, while many novel synthetic and semi-synthetic opioids were ultimately developed (including those acting at *kappa* and *delta* opioid receptors as well), the search for this perfect analgesic agent without abuse potential continues to this day.

In the 1990s, a new opioid addiction epidemic emerged in the USA, but this time, it was the consequence of widespread misuse and abuse of opioids available only through doctors' prescriptions, rather than due to illicit opium and heroin abuse (Compton and Volkow 2006). The start of this still ongoing epidemic was attributable to multiple factors, including aggressive marketing by pharmaceutical companies launching new sustained-release formulations with the claims that they were not addictive due to the slowed exposure (Van Zee 2009) along with increased pressure on physicians to treat pain (whereas past years saw an aversion to opioid prescribing for fear of drug addiction leaving patients often with untreated pain) both from professional societies and from medical oversight agencies. Thus, opioid prescribing increased dramatically with physicians writing long-term prescriptions for all types of chronic non-cancer pain conditions not previously treated with opioids (i.e., lower back pain, knee pain). While the efficacy of opioids is still not established for chronic non-cancer pain conditions, the harms of chronic opioids clearly are (Chou et al. 2015). Unfortunately, while the USA may have been ground zero for the prescription opioid epidemic, other countries, now more than a decade later, are experiencing similar problems with rising misuse, abuse, and other adverse medical consequences of prescribed opioids, including Western Europe, Canada, and Australia (e.g., see Blanch et al. 2014; Casati et al. 2012; Fischer et al. 2015). Numerous changes to regulatory control and practice procedures in the USA, including prescription monitoring programs, improved physician education, the Food and Drug Administration requirement of class-wide Risk Evaluation and Mitigation Strategies (FDA 2009), and changes to practice guidelines (CDC 2016), have been implemented with the aim of curbing this epidemic. This newest epidemic has once again brought the urgency of identifying effective analgesics *without* abuse potential or analgesics formulated to protect against abuse to the foreground and has also led to increased attention to abuse potential evaluations and improving experimental methodologies to assess these novel formulations.

2 Abuse Potential Evaluation: Methods

Abuse potential and abuse liability are often used interchangeably; however, they are actually different concepts. Abuse potential is an assessment of the likelihood that a particular drug will be abused based upon controlled data collected in a laboratory setting using accepted and (often) validated procedures. Abuse liability refers to a broader range of factors that will influence the likelihood of drug abuse in a societal context. For example, this may include drug availability and observed rates of abuse of related compounds, in addition to the pharmacological properties of the agent. Thus, abuse potential is one component of abuse liability. The full suite of abuse potential assessments of a novel agent prior to regulatory approval and marketing has a number of components. These include preclinical studies of (1) drug self-administration (to directly assess reinforcing efficacy and likelihood of abuse), (2) drug discrimination (to identify how similar the test drug is to the prototypes in its pharmacological class and/or to identify which pharmacological class a novel entity most closely resembles with regard to its CNS activity), and (3) dependence potential studies, which examine whether chronic exposure to the drug leads to an observable withdrawal syndrome following abrupt cessation. Once completed, abuse potential assessment studies are then conducted in humans.

Dependence potential studies, while historically conducted in humans, were discontinued for ethical and safety reasons. Moreover, while drug discrimination and self-administration procedures have been translated into the human laboratory and can be conducted, these are neither the typical nor the FDA-mandated approach to abuse potential assessment in humans. Rather, abuse potential studies focus on evaluating the effects of acute doses of the test drug in individuals who actively abuse opioids but who are not presently physically dependent on opioids with test doses spanning the therapeutic to the supratherapeutic range. Studies in normal healthy volunteers (i.e., without opioid abuse histories) are sometimes conducted, but these are not considered to meet the definition of abuse potential testing. It is worth noting that, while the abuse potential of opioids is quite high, many opioid-naïve individuals find them to be quite dysphoric because of their side effects, including pruritus, nausea, and vomiting. Opioid-experienced individuals are enrolled because, by their own history, they are sensitive to the euphoric effects of opioids and reflect the population for whom the risk of abuse is a concern. These studies always include a placebo control and a positive control condition (e.g., morphine) known to produce a signal. When the positive control condition fails to produce a signal on key outcomes that is significantly different from placebo, this is considered a failed study. To avoid this, it is common to conduct a qualification phase prior to randomization into a study; during this phase, subjects are exposed to the positive control and placebo on separate test occasions. Predetermined criteria for scores and responses must be met in order to qualify for enrollment. These studies are also conducted under double-blind conditions and often confine subjects as inpatients for the duration of testing to preclude the use of illicit drugs.

Many of the commonly employed outcome measures for abuse potential studies were developed during the golden era of addiction research at the Public Health Addiction Research Center, including the Addiction Research Center Inventory (ARCI; Haertzen 1974) and the Pharmacological Class Questionnaire (Fraser et al. 1961). The visual analog scales are used to assess subjective experiences of "drug liking," "high," "good effects," and "bad effects" to capture the time–action and magnitude of the drug response. Adjective checklists, often using a Likert scale, are used to assess specific symptoms and signs by the participant and blinded observer, respectively. Another commonly used measure is the Profile of Mood States (POMS; McNair et al. 1971). Frequently, physiological measures for safety (e.g., oxygen saturation or respiratory rate) and objective assessment of opioid response (e.g., pupil diameter) are collected concurrently in these studies. Other reviews provide more detailed information on study design and outcome measures for the interested reader (Balster and Bigelow 2003; Carter and Griffiths 2009; Preston and Walsh 1998; Romach et al. 2014). Abuse potential evaluation is now a requisite in the USA at least for approval of new drugs that have CNS activity and serve an important role in determining the regulatory control status over the drug (in the USA, this determines schedule placement within the Controlled Substances Act); thus, the methods are more broadly applied to pharmacological classes in addition to opioids (FDA 2010). However, the focus of this review will be on human laboratory evaluations of opioids that have reached clinical testing and/or are marketed at least somewhere in the world for a medical indication.

Recently, new methods have been added to the traditional approach of abuse potential evaluation in order to characterize the newly emerging classes of abuse-deterrent products (FDA 2015). Abuse-deterrent formulations (ADFs) are designed to deter abuse or misuse (e.g., by an alternative route), often of existing marketed drugs (especially opioids), either by mechanical, chemical, or other means. This chapter closes with a brief description and update on abuse potential methods and their application to the evaluation of abuse-deterrent opioid products.

3 Full *Mu* Opioid Agonists

Full *mu* opioid agonists are drugs that bind to the *mu* opioid receptor site and exert maximal agonist-like effects when bound to the receptor (e.g., increasing doses lead to increased effects, until maximal effects are reached). This action at *mu* opioid receptors produces an array of dose-dependent physiological effects that can be both beneficial and detrimental that include analgesia, cough suppression, constipation, miosis, pruritus, urinary retention, nausea, and decreased respiration (and respiratory failure at sufficiently high doses). Full *mu* opioid agonists also produce euphorigenic effects that can lead to abuse and, after repeated administration, a robust physical dependence syndrome.

Most prescription opioid analgesics (with some noted exceptions, as described in later sections) are full *mu* opioid agonists (see Table 1), as is heroin. This shared

Table 1 Commonly prescribed opioid compounds, example trade names and representative abuse potential studies

Opioid compounds	Immediate-release oral formulations		Extended-release oral formulations	Parenteral formulations	Representative abuse potential studies	
Full mu opioid agonists						
Oxycodone	Roxicodone™ +Oxaydo® *Endocet® *Percocet® *Endodan®	*Roxicet™ (tablets and oral solution) OxyNorm® (oral solution, tablets) Endone® *Percodan®	+OxyContin *Xartemis XR™ +Targin	Proladone (suppositories) Oxynorm® (suppositories)	Walsh et al. (2008b) Stoops et al. (2010) Comer et al. (2010) Lofwall et al. (2012) Middleton et al. (2012)	Schoedel et al. (2012) Walsh et al. (2012) Colucci et al. (2014) Harris et al. (2014)
Hydrocodone	*Lorcet® Lortab® *Lortab Elixir® (oral solution) *Vicodin® *Vicoprofen® *Zydone® *Norco®	*Maxidone® Hycet® (oral solution) Hycodan® (oral solution, tablets) Liquicet™ (oral solution)	+Hysingla ER®	–	Kaplan et al. (1997) Zacny et al. (2005), Zacny and Gutierrez (2009) Walsh et al. (2008b)	
Morphine	Ordine® (oral solution) Sevredol® Anamorph®		^Embeda® Kadian® MS Contin® Kapanol® MS Mono® Momex SR®	Infumorph® Duramorph® (injection formulations) DepoDur® (depot injection)	Eddy and Lee (1959) Fraser et al. (1978), Jasinski and Preston (1986)	Walker and Zacny (1999) Heishman et al. (2000) Marsch et al. (2001)

(continued)

Table 1 (continued)

Opioid compounds	Immediate-release oral formulations	Extended-release oral formulations	Parenteral formulations	Representative abuse potential studies	
				Lamb et al. (1991) Preston et al. (1991) Lamas et al. (1994)	Zacny and Lichtor (2008) Comer et al. (2008b) Stoops et al. (2010)
Oxymorphone	Opana®	+Opana ER®	Numorphan® (injection, suppositories)	Eddy et al. (1959) Eddy and Lee (1959) Schoedel et al. (2011) Babalonis et al. (2014)	
Hydromorphone	Dilaudid® (oral liquid, oral tablets)	Jurnista® Exalgo®	Dilaudid® Dilaudid-HP® (injection formulations)	Bickel et al. (1986) Hill and Zacny (2000) Abreu et al. (2001) Walsh et al. (2008b)	
Fentanyl	Actiq® (transmucosal lozenge) Abstral® (sublingual tablet) Onsolis® (buccal film) Fentora® (buccal tablet) Subsys® (sublingual spray)		Duragesic® (transdermal patch) Sublimaze® (injection)	Zacny et al. (1992, 1996) Greenwald et al. (1996) Baylon et al. (2000) Comer et al. (2008b)	
Alfentanil			Alfenta® (injection)	Greenwald et al. (1996) Tompkins et al. (2014)	
Meperidine	Demerol®		Demerol® (injection)	Fraser et al. (1978) Zacny et al. (1993) Walker and Zacny (1999)	
Methadone	Dolophine® Physeptone® Biodone® (liquid)		Physeptone® (injection)	Isbell and Vogel (1949) Isbell et al. (1947)	Jasinski and Preston (1986)

(continued)

Table 1 (continued)

Opioid compounds	Immediate-release oral formulations	Extended-release oral formulations	Parenteral formulations	Representative abuse potential studies	
	Diskets® Methadose™ (oral liquid, tablets)		Generic methadone (injection)	Wikler et al. (1953) Fraser et al. (1978)	Greenwald (2008) Donny et al. (2005) Lintzeris et al. (2007)
Heroin	Prescription-grade diacetylmorphine (oral capsules)		Prescription-grade diacetylmorphine (IV)	Jasinski and Preston (1986) Comer et al. (1999, 2008b) Klous et al. (2005) Sullivan et al. (2006) Reviews as an opioid maintenance therapy, see: Fischer et al. (2002, 2007)	
Partial agonist					
Buprenorphine	^Suboxone® (sublingual film) ^Zubsolv® (sublingual tablet) Generic buprenorphine (sublingual tablet) ^Bunavail (buccal film) Belbuca (buccal film)		Butrans® (transdermal patch) Buprenex® (injection) Norspan® (transdermal patch)	Jasinski et al. (1978) Jasinski et al. (1989) Pickworth et al. (1993) Walsh et al. (1995a), b, Walsh et al. (1994) Strain et al. (2000)	Comer et al. (2002) Umbricht et al. (2004) Comer et al. (2005) Duke et al. (2010) Middleton et al. (2011) Weinhold et al. (1992)

(continued)

Table 1 (continued)

Opioid compounds	Immediate-release oral formulations	Extended-release oral formulations	Parenteral formulations	Representative abuse potential studies
Mixed opioid agonists/antagonists				
Pentazocine	^Talwin NX®		Talwin® (injection)	Fraser and Rosenberg (1964) Jasinski et al. (1970) Jasinski and Nutt (1972) Eddy (1973) Bickel et al. (1986) Preston et al. (1987) Preston and Bigelow (1993)
Nalbuphine	–	–	Nubain® (injection) Generic nalbuphine (injection)	Jasinski and Mansky (1972) Jasinski and Nutt (1972) Eddy (1973) Preston et al. (1989a) Preston et al. (1990)
Butorphanol	Generic oral tablet, nasal spray	–	Stadol® (injection) Generic product (injection)	Jasinski et al. (1975) Preston et al. (1988) Zacny et al. (1994) Greenwald and Stitzer (1998) Walsh et al. (2001) Walsh et al. (2008a)
Atypical opioids				
Tramadol	Tramal® (liquid and tablet formulations) Zydol® Tramedo® Ultram® *Ultracet®	Lodam SR Zydol SR Tramal SR® Tramedo SR® Ultram ER®	Tramal® (injection)	Preston et al. (1991) Jasinski et al. (1993) Epstein et al. (2006) Duke et al. (2011) Stoops et al. (2012, 2013) Babalonis et al. (2013)

(continued)

Table 1 (continued)

Opioid compounds	Immediate-release oral formulations		Extended-release oral formulations	Parenteral formulations	Representative abuse potential studies
Tapentadol	Nucynta®		Palexia SR® Nucynta ER®	–	
Codeine	*Nurofen Plus® *Aspalgin® *Prodeinextra® *Codalgin® *Codapane Forte® **Cheratussin AC® (oral solution) **Promethazine with codeine (oral solution)	*Comfarol Fort® *Prodeine Forte® *Panadeine Forte® *Tylenol® with codeine	–	–	Eddy et al. (1968) Walker and Zacny (1998) Kim et al. (2002) Babalonis et al. (2013)

List of trade names are examples and do not represent an exhaustive list
*Combination product (NSAID, ibuprofen, aspirin)
**Combination product (decongestant, antihistamine/anti-nausea)
^Product formulated with opioid antagonist (naloxone, naltrexone)
+Product contains novel mechanical/physical/chemical properties purported to be abuse deterrent

mechanism of action is also indicative of similar abuse potential profiles across this drug class, as these compounds produce similar and rather reliable dose-related increases in abuse potential measures (compared to placebo) in participants with opioid abuse histories. Many studies have been conducted exploring the abuse potential of this drug class (see Table 1 for references); it is not feasible to describe a full review of those studies here. Because most full *mu* opioids produce a similar constellation of abuse-related effects, oxycodone will be used to illustrate the prototypical profile of effects. Oxycodone was selected because it is commonly prescribed, frequently abused, has been widely investigated in the human laboratory, with reliable and replicated results, and produces effects similar to other full *mu* opioid agonists, including morphine, which was the comparator opioid used in numerous early studies at the Addiction Research Center (Eddy 1956).

Oxycodone, when administered via oral (Comer et al. 2010; Walsh et al. 2008b; Webster et al. 2012), intranasal (Harris et al. 2014; Middleton et al. 2012; Schoedel et al. 2012, Lofwall et al. 2012), subcutaneous (Eddy 1956), or intravenous (Colucci et al. 2014; Stoops et al. 2010) routes, produces dose-related increases in ratings of subjective drug effects indicative of abuse liability (relative to placebo). These measures include participants' ratings of the following queries: "Do you feel any drug effect?", "How much do you like the drug effect?", "How high are you?", and "Does the drug have any good effects?" (Harris et al. 2014; Walsh et al. 2008b). Oxycodone also produces increases in participant-rated adjective items, such as itchy skin, nodding, relaxed, talkative, dry mouth, and good mood, and increased ratings on the Morphine-Benzedrine Group (MBG) scale of the ARCI, reflecting euphoria (Stoops et al. 2010; Walsh et al. 2008b). Dose-related increases of street value estimates are also observed. This measure requires the participant to estimate how much they would pay for the blinded drug if given the opportunity to purchase it on the street. The street values obtained often roughly estimate the price of illicitly purchased drug in the participant's geographic region (e.g., oxycodone estimates are approximately $1/mg of drug, which closely resembles the street price) (Babalonis et al. 2014; Middleton et al. 2012).

Trained research staff observers, who are blind to dosing conditions, assess signs of opioid agonist-like effects and rate their magnitude in regular intervals throughout the study session. The array of effects that emerge after oxycodone administration is often similar to the items endorsed by participants and include increases in nodding, itchy skin, talkativeness, drunken, sluggish, and good mood; these observations generally follow the same time –action curve as the subject-rated effects and are dose-dependent, with higher composite scores reported when higher doses are administered (Walsh et al. 2012, 2008b).

Although not required (and not always included) in clinical abuse potential assessments, self-administration and drug choice procedures are sometimes included in the battery of measures collected. In these paradigms, participants (1) have the opportunity to work (e.g., repeatedly clicking a computer mouse) for none, some, or all of a previously sampled drug dose or (2) make a choice to elect to receive money or drug. Several studies have demonstrated that participants elect to work to obtain oxycodone and will forego money in order to receive an oxycodone dose (Babalonis et al. 2013; Comer et al. 2010; Middleton et al. 2012) relative to placebo. With drug identification measures, oxycodone and other full *mu* opioid agonists are readily discriminated from other drug classes (e.g., stimulants such as cocaine/amphetamine; benzodiazepines such as alprazolam) when queried (Jasinski 1977).

Physiological effects of full *mu* opioid agonists are also measured during abuse potential testing. Oxycodone and other full *mu* opioid agonists produce dose-dependent effects on miosis (e.g., pupil constriction, often measured using a pupillometer or photograph) and respiratory depression (measured as respiratory rate, oxygen saturation, or end tidal carbon monoxide levels) (Fraser et al. 1976; Stoops et al. 2010). Overall, this array of participant-rated, observer-rated, behavioral, and physiological outcome measures is generally internally consistent (e.g., rather good concordance between the abuse potential measures) and often corresponds to both animal studies of abuse potential (e.g., self-administration) and epidemiological studies (Comer et al. 2008a, 2012). When individual short-acting opioid agonists are compared within the same study, the individual drug profiles are often comparable when equipotent doses are selected. For example, oral doses of oxycodone (10, 20, 40 mg), hydromorphone (10, 17.5, 25 mg), hydrocodone (15, 30, 45 mg), and placebo were tested using a randomized, within-subject design to assess their relative abuse potential in non-dependent opioid abusers (Walsh et al. 2008b). The results indicated that oxycodone and hydrocodone were approximately equipotent across the array of subjective, observer-rated, and physiological measures collected. Hydromorphone was modestly more potent (approximately twofold) than oxycodone (although they differ more considerably in their analgesic potency, Jaffe and Martin 1991); however, all drugs displayed a similar time course (peak effects, duration of physiological effects, subjective effects), and all were qualitatively similar in regard to abuse potential profiles.

There are instances when between-drug comparisons of opioid agonists reveal differences between drugs; however, these differences are typically related to dose/potency differences. Ideally, test doses should be equipotent and produce the same maximal effect on a selected measure (equal effects on miosis are the standard) so that accurate between-drug comparisons can be achieved. Other factors that can influence the magnitude of effect produced by a full *mu* agonist include route of administration (e.g., oral vs. IV administration), rate of drug onset [which can be influenced by drug formulation differences (immediate-release, extended-release)], and bioavailability. These factors, along with the half-life of the compound, can affect the time course and duration of effects. Most immediate-release oral opioids display half-lives that are approximately 3–4 h, producing pronounced subjective effects approximately 1–4 h post-dose and physiological effects (miosis, decreased O_2 saturation and respiratory rate) from 1–5 + h post-dose; however, very short-acting (fentanyl) and long-acting (methadone) agonists produce unique time course profiles based on their pharmacological half-lives.

The population in which the drug is tested also influences the profile of effects. For example, genetic and physiological differences, particularly in liver enzyme CYP 450 function (e.g., rapid or poor metabolizers), are present in a subset of the population, and these differences can impact the abuse liability profile of drugs that rely on hepatic metabolism (Kathiramalainathan et al. 2000; Otton et al. 1993). Further, participants' opioid use histories can also influence outcome measures. Participants with little to no experience with opioids cannot tolerate a supratherapeutic dose range (which is recommended for abuse potential studies); thus, studies with naïve/light users are limited to a therapeutic dose range. However, these participants are not as likely to report abuse-related, euphoric-like effects and often report aversive, dysphoric effects (nausea, dizziness) (Comer et al. 2010; Zacny and Gutierrez 2009; Zacny and Lichtor 2008). Conversely, participants who are physically dependent (and thus, quite tolerant) will often require much higher doses (higher even than the supratherapeutic doses tested in non-dependent abusers) to produce euphorigenic effects; however, the overall abuse potential profile of the drug is similar to that reported by experienced non-dependent participants (Comer et al. 2013).

4 *Mu* Opioid Partial Agonists

4.1 Buprenorphine: A Case Study

Buprenorphine is a semi-synthetic opioid derived from thebaine (Lewis 1974) that is widely used for the treatment of pain and opioid dependence. Buprenorphine is a *mu* partial agonist (Cowan et al. 1977), and while it possesses other pharmacological activity (kappa antagonist and partial agonist at the nociceptin receptor), these other activities are not typically considered to be of relevance when used in clinical practice. Therefore, buprenorphine will be described here as an exemplar of a *mu* partial opioid agonist as substantial research has been conducted with it under a very broad array of clinical conditions yielding a rich database.

In individuals who are opioid abusers but are not physically dependent on opioids, a partial *mu* agonist would be predicted to be similar or identical to a full agonist in an abuse potential assessment but only over some portion of the dose–response curve. It would also be expected that the efficacy would plateau at some dose with higher doses producing no greater effect; thus, a partial agonist would ultimately possess lower intrinsic activity compared to a full agonist after that plateau is reached. In other words, the partial agonist will produce less effect than a full agonist, but at what dose range this occurs will determine whether or not this plateau or ceiling effect is clinically relevant or evident. With buprenorphine, early studies in humans suggested that it would have reduced abuse potential compared to full opioid agonists (Jasinski et al. 1978). However, subsequent studies revealed that while the maximum acute effects on abuse potential measures were achieved at approximately 16 mg sublingual buprenorphine (Walsh et al. 1995b, 1994) with the dose–response curve relatively flat thereafter (seen similarly with intravenous dosing in Umbricht et al. 2004), these effects were similar in magnitude to those produced by an acute dose of 60 mg methadone (p.o.), which is a very substantial dose of methadone in a non-dependent individual. These data suggested that while buprenorphine may have somewhat diminished abuse potential compared to a full agonist, overall it was still associated with substantial abuse liability. This fact has been borne out following the marketing of buprenorphine products for the treatment of opioid dependence and the significant misuse and diversion of the product that followed (for recent review see Lofwall and Walsh 2014).

Numerous studies have been conducted to examine the abuse potential of buprenorphine in opioid-experienced subjects without physical dependence and have reported significant and dose-dependent increases on measures of abuse-related subjective effects (i.e., ratings of "drug liking," street value, and measures of euphoria). This has been reported for buprenorphine given by the sublingual (Duke et al. 2010; Strain et al. 2000; Walsh et al. 1995b), intravenous (Comer et al. 2002; Pickworth et al. 1993), subcutaneous (Jasinski et al. 1989), intranasal (Middleton et al. 2011), and intramuscular routes of administration (Duke et al. 2010; Weinhold et al. 1992); buprenorphine has poor oral bioavailability, and thus, few studies have examined this route. In a study by Comer and colleagues (Comer et al. 2005) that compared the abuse potential of intravenous buprenorphine and methadone in recently detoxified heroin-dependent subjects, it was found that the two drugs were indistinguishable on subjective measures of "good drug effects," "liking," and other typical abuse potential outcomes and concluded that they were comparable with regard to both subjective outcomes and reinforcing efficacy as measured by self-administration.

In individuals who are physically dependent on opioids, the abuse liability of buprenorphine may differ considerably from that of a full *mu* opioid agonist. That is, because of its lower intrinsic activity, buprenorphine itself can precipitate opioid withdrawal by displacing a full agonist (i.e., with higher intrinsic activity) from the receptor. Buprenorphine-precipitated withdrawal has been demonstrated in individuals maintained on methadone doses ranging from 30 to 60 mg (Strain et al. 1992, 1995; Walsh et al. 1995a). These studies have shown that the magnitude of precipitated withdrawal increases with increasing methadone maintenance dose and also with increasing doses of buprenorphine. In contrast, studies conducted in subjects maintained on shorter-acting opioids, such as morphine (Comer et al. 2008b; Mendelson et al. 1999; Schuh et al. 1996), have reported no buprenorphine-precipitated withdrawal, but this may be attributable to maintenance dose and timing of the challenge because case reports (see, e.g., Clark et al. 2002) and clinical experience have found buprenorphine-precipitated withdrawal in individuals dependent on short-acting opioids, such as heroin. Thus, while the risk of precipitating withdrawal by taking buprenorphine may be viewed as a form of abuse deterrence, in clinical practice it is desirable to avoid the risk of precipitated withdrawal in patients who are attempting to initiate treatment with buprenorphine for opioid dependence. Overall, the likelihood of buprenorphine-precipitated withdrawal can be decreased by using an initial low dose (Rosado et al. 2007) and by administering it when the individual is already in a state of mild withdrawal from their opioid of choice. Indeed, this has become the recommended procedure for inducting patients onto buprenorphine maintenance for the treatment of opioid dependence (CSAT 2004). Conversely, several studies have examined the effects of buprenorphine in individuals maintained on buprenorphine itself (Jones et al. 2014, 2015; Strain et al. 1997). In these instances, buprenorphine does not precipitate frank opioid withdrawal (as would be expected based on the underlying pharmacological principles related to receptor affinity).

5 Mixed Opioid Agonists/Antagonists

Opioid compounds believed to exert their effects simultaneously through more than one opioid receptor at clinically relevant doses are often referred to as mixed opioid agonists/antagonists. The pharmacological profile of these agents in vivo does not exhibit a profile of action that can be attributed to single receptor activation or blockade. Numerous compounds fitting this classification were synthesized and tested during the period of the early federally directed US research program (Eddy 1973). A few mixed opioid agonists/antagonists are marketed for their analgesic properties, and this discussion will focus on three drugs that fall into this category: pentazocine, butorphanol, and nalbuphine.

Pentazocine was characterized as an antagonist or partial agonist at the *mu* receptor and as a *kappa* agonist through early preclinical studies (Martin 1983). The clinical findings generally support this characterization. When acute doses of pentazocine are administered to non-dependent individuals, it produces a profile of subjective effects similar to *mu* opioid agonists, including elevated ratings of drug liking, and increases on the MBG scale of the ARCI (sensitive to morphine-like euphoric effects) (Preston et al. 1987, 1992), but tends to be weaker than a full agonist (Fraser and Rosenberg 1964). However, as higher doses are given, the profile of effects is modified with subjective ratings increasing on measures reflecting dysphoria, such as "nervous" and "bad effects" (Jasinski et al. 1970; Preston et al. 1987), consistent with *kappa* agonist actions. Moreover, the *mu* and *kappa* effects have been dissociated through the use of naltrexone blockade to selectively block the *mu*-mediated effects (Preston and Bigelow 1993). Although no full kappa agonists are marketed, several have made it to clinical testing and are generally quite dysphoric (including feelings of depersonalization and perceptual disturbances) and sedating (see, e.g., Walsh et al. 2001).

Nalbuphine is similar to pentazocine as its effects are mediated through both the *mu* and *kappa* receptors, where it may function as a partial agonist at both. At lower doses, nalbuphine appears to be morphine-like (Jasinski and Mansky 1972), but with higher doses, the profile changes and includes increased ratings of sedation, "drunk," and measures related to dysphoria (the LSD scale of the ARCI) (Jasinski and Mansky 1972; Preston and Bigelow 1994; Preston et al. 1989a). As with buprenorphine, pentazocine and nalbuphine have been shown to produce precipitated withdrawal in opioid-dependent individuals, supporting their characterization as *mu* partial agonists. Specifically, early studies conducted at the ARC aimed at attempting to characterize the pharmacological profile of these then novel opioids reported that both pentazocine- and nalbuphine-precipitated withdrawal in individuals maintained on 240 and 60 mg of morphine/day (i.m.), respectively (Jasinski and Nutt 1972). Subsequent studies reported that pentazocine (Lamas et al. 1994; Strain et al. 1993) and nalbuphine (Preston et al. 1989b) could also precipitate withdrawal in individuals maintained on methadone.

Butorphanol is characterized as having intermediate/partial agonist activity at *mu*, *kappa*, and *delta* opioid receptors (Commiskey et al. 2005; Horan and Ho 1989; Pircio et al. 1976). Acute doses of butorphanol produce a constellation of subjective effects that appears to be the result of both *mu* activation (euphoria, modest "liking" scores) and *kappa* activation (increased sedation, difficulty concentration and "floating") (Greenwald and Stitzer 1998; Jasinski et al. 1975; Walsh et al. 2008a; Zacny et al. 1994). Moreover, butorphanol can produce physiological responses consistent with kappa agonist activity, including sweating and diuresis, the latter when given in combination with a *mu*-blocking dose of naltrexone (Walsh et al. 2008a; Zacny et al. 1994). Similar to the others though, butorphanol too can precipitate opioid withdrawal when given at sufficiently high doses to methadone-dependent individuals (Preston et al. 1988). Thus, butorphanol, like pentazocine and nalbuphine, appears to behave as a *mu* partial agonist given its capacity to produce *mu*-like effects when given acutely and to precipitate withdrawal in opioid-dependent individuals; however, it may also have more robust *kappa* agonist effects compared to the other marketed mixed agonist/antagonists. Consistent with the laboratory findings for this class of agents, the mixed agonist/antagonists are less regulated compared to the full mu agonists (all of which are in Schedule II in the USA), with butorphanol and pentazocine in Schedule IV and nalbuphine unscheduled in the USA.

6 Atypical Opioid Agonists

Atypical opioids are drugs that have some action at the *mu* opioid receptor, but are either (1) pro-drugs—a parent drug that has little to no action at the *mu* opioid receptor until is metabolized to an active opioid constituent or (2) dual agonists—drugs that act as agonists at the *mu* opioid receptor and at least one other receptor system (e.g., *mu* opioid and norepinephrine agonist). The most commonly prescribed atypical opioid is codeine, which is available in many countries (marketed as an over-the-counter medication in Canada, Australia, and many European countries, while it remains a prescription drug in the USA). The parent compound, codeine, displays minimal affinity at the *mu* opioid receptor and does not display a great degree of analgesic efficacy once bound to the receptor (Volpe et al. 2011). However, codeine is hepatically metabolized by enzyme CYP P450 2D6 into morphine, a full *mu* opioid agonist. Thus, adequate 2D6 activity is necessary for codeine to produce its physiological and subjective effects, and this hepatic activity also modulates its abuse liability (Kathiramalainathan et al. 2000). In studies that have directly examined the abuse potential of codeine, the results have been generally consistent—a supratherapeutic dose of codeine (200 mg) produces moderate signals for abuse potential (e.g., "drug liking," "high") and modest respiratory depression; however, these effects were lesser than those observed when directly compared with oxycodone (Babalonis et al. 2013). Nonetheless, codeine abuse is common in several countries where it is readily available or when access to full *mu* agonists is limited (McAvoy et al. 2011; Peters et al. 2007 and see review in this volume by Nielsen). Codeine can also be synthetically manipulated to create desomorphine (e.g., the main constituent in "krokodil"), a full *mu* opioid agonist. Desomorphine abuse has been most frequently reported in Russia and neighboring Soviet Republics, but cases in several European countries, the USA and Australia, have also recently emerged as a heroin substitute (Alves et al. 2015).

Tramadol is also a pro-drug, with the parent compound (tramadol) acting as a serotonin and norepinephrine reuptake inhibitor, and its active hepatic metabolite *o*-desmethyltramadol (also known as M1) resulting from tramadol metabolism by CYP 2D6 is a moderate efficacy *mu* opioid agonist. Several studies have examined the abuse potential of parenteral tramadol (e.g., intramuscular and intravenous administration) (Preston et al. 1991). High doses of IV tramadol (300, 700 mg) produced seizures (likely due to profound serotonergic activity), while lower IV doses (100, 200 mg) were safely tolerated (Epstein et al. 2006; Jasinski et al. 1993) but produced placebo-like ratings on subjective measures of abuse potential compared to morphine. Intramuscular tramadol (75, 150, 300 mg) produced a similar profile of abuse-related effects in which lower doses were placebo-like, while the highest dose (300 mg, IM) produced relatively modest increases in drug effects and virtually no miosis, compared to the morphine control (Epstein et al. 2006; Preston et al. 1991). This led to the conclusion that tramadol had limited to no abuse potential. However, oral administration of tramadol (the route by which it is most commonly taken clinically), which produces greater concentrations of the opioid metabolite (M1) relative to parenteral administration (which largely bypasses hepatic metabolism) (Ardakani and Rouini 2007; Enggaard et al. 2006), displays a composite profile of greater abuse potential—this is in contrast to other opioids, which produce greater euphoric effects with parenteral administration. For example, oral doses of tramadol (200, 400 mg) produce respiratory depression similar to oral oxycodone (20, 40 mg), and higher doses of oral tramadol (400 mg) are self-administered to a similar degree to 40 mg of oxycodone (Babalonis et al. 2013). Oral tramadol (200, 400 mg) was also identified as opioid agonist in a drug discrimination paradigm (Duke et al. 2011); however, *mu* opioid agonist effects do not appear to be the only component driving the abuse-related profile of tramadol, as the opioid antagonist naltrexone only partially blocks the subject-rated abuse profile, suggesting that some of the abuse liability of tramadol is driven by monoaminergic effects (Stoops et al. 2012).

Tapentadol is an atypical opioid with mixed pharmacological action—the parent compound acts as an agonist at *mu* opioid receptors and blocks the reuptake of norepinephrine (e.g., a dual action agonist). Although tapentadol is marketed in Australia, the USA, and several European countries, there have not been any published studies examining its abuse potential. One study examined the effects of therapeutic doses of oral tapentadol (25, 50, 75 mg) compared to tramadol (50, 100, 150 mg) and hydromorphone (2, 4, 6 mg) in a sample of occasional opioid users (Stoops et al. 2013). The results indicated that the high dose of tapentadol (75 mg) produced similar outcomes as the moderate and high doses of tramadol (100, 150 mg) and hydromorphone (4, 6 mg) on peak miosis and several subject-rated measures (e.g., street value estimates, good drug effects). However, additional abuse potential studies examining supratherapeutic doses are necessary to further understand the characteristics of tapentadol.

6.1 Abuse-Deterrent Formulations (ADFs) and Abuse Potential Testing

The US Food and Drug Administration has developed guidance on the approach to assessing abuse potential for ADFs (FDA 2015). These new methods largely focus on assessing the degree of tamper resistance for formulations with physical/chemical barriers designed mechanically to resist crushing and decrease solubility. For example, various strategies are used to create hardened oral tablets that are very difficult to crush, thereby reducing the likelihood that a tablet will be crushed and used by the intranasal or intravenous route. These technologies have been primarily applied to extended-release opioid formulations as they are intended to deliver continuous drug over a protracted period and, thus, contain a higher dosage than many immediate-release opioids (e.g., the reformulated version of Oxycontin® PurduePharma 2015); however, there is more recent interest in applying these technologies to immediate-release opioids. In these studies, various implements are employed to crush the opioid product and the resultant particle size is measured. The abuse potential of the crushed product versus intact product may be evaluated in a human abuse potential study (Harris et al. 2014). Additives may be included in some formulations that prevent the broken particles from dissolving easily through the addition of gelling agents; thus, when water or another solution is used in an attempt to dissolve the drug, the mixture gels into a viscous goop rendering it difficult/impossible to inject as a solution. In this case, the viscosity of the resulting product after dissolution in alcohol, water, common beverages, or common chemicals is evaluated and the "syringeability" of the solution is then measured (but not injected in humans).

Other strategies that are employed to deter abuse include the addition of an opioid antagonist to precipitate opioid withdrawal when misused (as with Suboxone®, a formulation combining buprenorphine and naloxone in a 4:1 ratio), or the addition of an aversive agent, such as niacin (Acurox®) or sodium lauryl sulfate (Oxecta®; Schoedel et al. 2012), which when absorbed, can cause unpleasant flushing reactions or burning sensations, respectively. These types of deterrent strategies require additional human studies to characterize the extent of the aversive effects and conditions under which they occur (e.g., by route of administration, dependence status of the individual). Depot formulations and implant technologies are yet another approach to abuse deterrence that are being widely explored. The methodology for evaluating abuse potential of abuse-deterrent agents is still evolving and will continue to do so as additional inventive approaches to deterrence are developed.

7 Summary

There is a rich and lengthy history of abuse potential testing that emanated from a public health imperative to identify novel agents that could be designed to have great clinical efficacy as analgesics but without the associated risks of abuse, misuse, and dependence. This initiative, launched nearly a century ago, was triggered by an epidemic of opium addiction. It is ironic (and unfortunate) that now renewed efforts, interest, and resources are being directed to increasing scientific activity in this research area because of yet another opioid abuse epidemic—this one triggered by abuse and misuse of licit opioids. Successful efforts early on led to the development of validated and accepted scientific methodologies and a framework to assess abuse potential of novel agents. Indeed, novel agents were developed, including the mixed opioid agonist/antagonists, some of which made it to the marketplace and are associated with lower abuse liability (although sometimes with decreased analgesic efficacy and additional unpleasant side effects). Great advances in understanding the molecular neuropharmacology of the opioid receptor systems have been made in more recent years that are now driving the development of the next generation of novel entities (for example, Soergel et al. 2014; Toll et al. 2016; White et al. 2015) as well as novel formulations of older drugs intended to deter abuse. Given the current numbers of individuals physically dependent on opioids and the development of agents that may be intended to deter abuse only in that population, additional improvements and guidance for methods are needed to address this gap. As the prescription opioid epidemic and high rates of fatal overdose do not appear to be waning in the USA and are simultaneously on the rise elsewhere in the world, laboratory evaluations of novel opioids and ADFs will remain a critical element of drug development.

References

Abreu ME, Bigelow GE, Fleisher L, Walsh SL (2001) Effect of intravenous injection speed on responses to cocaine and hydromorphone in humans. Psychopharmacology 154:76–84

Alves EA, Soares JX, Afonso CM, Grund JP, Agonia AS, Cravo SM, Netto AD, Carvalho F, Dinis-Oliveira RJ (2015) The harmful chemistry behind "krokodil": street-like synthesis and product analysis. Forensic Sci Int 257:76–82

Ardakani YH, Rouini MR (2007) Pharmacokinetics of tramadol and its three main metabolites in healthy male and female volunteers. Biopharm Drug Dispos 28:527–534

Babalonis S, Lofwall MR, Nuzzo PA, Siegel AJ, Walsh SL (2013) Abuse liability and reinforcing efficacy of oral tramadol in humans. Drug Alcohol Depend 129:116–124

Babalonis S, Lofwall MR, Nuzzo PA, Walsh SL (2014) Pharmacodynamic effects or oral oxymorphone: abuse liability, analgesic profile and direct physiologic effects in humans. Addict Biol e-pub ahead of print

Balster RL, Bigelow GE (2003) Guidelines and methodological reviews concerning drug abuse liability assessment. Drug Alcohol Depend 70:S13–S40

Baylon GJ, Kaplan HL, Somer G, Busto UE, Sellers EM (2000) Comparative abuse liability of intravenously administered remifentanil and fentanyl. J Clin Psychopharmacol 20:597–606

Bickel WK, Preston KL, Bigelow GE, Liebson IA (1986) Three-way drug discrimination in post-addict volunteers: hydromorphone, pentazocine, and saline. NIDA Res Monogr 67:177–183

Blanch B, Pearson SA, Haber PS (2014) An overview of the patterns of prescription opioid use, costs and related harms in Australia. Br J Clin Pharmacol 78:1159–1166

Campbell ND, Olsen JP, Walden L (2008) The narcotic farm the rise and fall of America's first prison for drug addicts. Abrams Books, New York

Carter LP, Griffiths RR (2009) Principles of laboratory assessment of drug abuse liability and implictations for clinical development. Drug Alcohol Depend (in press)

Casati A, Sedefov R, Pfeiffer-Gerschel T (2012) Misuse of medicines in the European Union: a systematic review of the literature. Eur Addict Res 18:228–245

CDC (2016) CDC guideline for prescribing opioids for chronic pain—United States, 2016. MMWR Morb Mortal Wkly Rep 65:1–49

Chou R, Turner JA, Devine EB, Hansen RN, Sullivan SD, Blazina I, Dana T, Bougatsos C, Reyo RA (2015) The effectiveness and risks of long-term opioid therapy for chronic pain: a systematic review for a National Institutes of Health Pathways to Prevention Workshop. Ann Intern Med 162:276–286

Clark NC, Lintzeris N, Muhleisen PJ (2002) Severe opiate withdrawal in a heroin user precipitated by a massive buprenorphine dose. Med J Aust 176:166–167

Colucci SV, Perrino PJ, Shram M, Bartlett C, Wang Y, Harris SC (2014) Abuse potential of intravenous oxycodone/naloxone solution in nondependent recreational drug users. Clin Drug Investig 34:421–429

Comer SD, Collins ED, MacArthur RB, Fischman MW (1999) Comparison of intravenous and intranasal heroin self-administration by morphine-maintained humans. Psychopharmacology 143:327–338

Comer SD, Collins ED, Fischman MW (2002) Intravenous buprenorphine self-administration by detoxified heroin abusers. J Pharmacol Exp Ther 301:266–276

Comer SD, Sullivan MA, Walker EA (2005) Comparison of intravenous buprenorphine and methadone self-administration by recently detoxified heroin-dependent individuals. J Pharmacol Exp Ther 315:1320–1330

Comer SD, Ashworth JB, Foltin RW, Johanson CE, Zacny JP, Walsh SL (2008a) The role of human drug self-administration procedures in the development of medications. Drug Alcohol Depend 96:1–15

Comer SD, Sullivan MA, Whittington RA, Vosberg SK, Kowalczyk WJ (2008b) Abuse liabliity of prescription opioids compared to heroin in morphine-maintained heroin abusers. Neuropsychopharmacology 33:1179–1191

Comer SD, Sullivan MA, Vosburg SK, Kowalczyk WJ, Houser J (2010) Abuse liability of oxycodone as a function of pain and drug use history. Drug Alcohol Depend 109:130–138

Comer SD, Zacny JP, Dworkin RH, Turk DC, Bigelow GE, Foltin RW, Jasinski DR, Sellers EM, Adams EH, Balster R, Burke LB, Cerny I, Colucci RD, Cone E, Cowan P, Farrar JT, Haddox JD, Haythornthwaite JA, Hertz S, Jay GW, Johanson CE, Junor R, Katz NP, Klein M, Kopecky EA, Leiderman DB, McDermott MP, O'Brien C, O'Connor AB, Palmer PP, Raja SN, Rappaport BA, Rauschkolb C, Rowbotham MC, Sampaio C, Setnik B, Sokolowska M,

Stauffer JW, Walsh SL (2012) Core outcome measures for opioid abuse liability laboratory assessment studies in humans: IMMPACT recommendations. Pain 153:2315–2324
Comer SD, Metz VE, Cooper ZD, Kowalczyk WJ, Jones JD, Sullivan MA, Manubay JM, Vosburg SK, Smith ME, Peyser D, Saccone PA (2013) Comparison of a drug versus money and drug versus drug self-administration choice procedure with oxycodone and morphine in opioid addicts. Behav Pharmacol 24:504–516
Commiskey S, Fan L-W, Ho IK, Rockhold RW (2005) Butorphanol: effects of a prototypical agonist-antagonist analgesic on k-opioid receptors. J Pharmacol Sci 98:109–116
Compton WM, Volkow ND (2006) Abuse of prescription drugs and the risk of addiction. Drug Alcohol Depend 83(Suppl 1):S4–S7
Cowan A, Lewis JW, Macfarlane IR (1977) Agonist and antagonist properties of buprenorphine, a new antinociceptive agent. Br J Pharmacol 60:537–545
CSAT (2004) Clinical guidelines for buprenorphine in the treatment of opioid addiction. Substance abuse and mental health services administration
de Quincy T (1823) Confessions of an English Opium-eater. Taylor and Hessey, London
Donny EC, Brasser SM, Bigelow GE, Stitzer ML, Walsh SL (2005) Methadone doses of 100 mg or greater are more effective than lower doses at suppressing heroin self-administration in opioid-dependent volunteers. Addiction 100:1496–1509
Duke AN, Correia CJ, Walsh SL, Bigelow GE, Strain EC (2010) Acute effects of intramuscular and sublingual buprenorphine and buprenorphine/naloxone in non-dependent opioid abusers. Psychopharmacology 211:303–312
Duke AN, Bigelow GE, Lanier RK, Strain EC (2011) Discriminative stimulus effects of tramadol in humans. J Pharmacol Exp Ther 338:255–262
Eddy NB (1956) Synthetic substances with morphine-like effect. Bull World Health Organ 114:353–402
Eddy NB (1973) The national research council involvement in the opiate problem. National Academy of Sciences, Washington, DC
Eddy NB, Lee LE Jr (1959) The analgesic equivalence to morphine and relative side action liability of oxymorphone (14-hydroxydihydro morphinone). J Pharmacol Exp Ther 125:116–121
Eddy NB, Lee LE Jr, Harris C (1959) Physical dependence and tolerance in relation to various analgesics in patients suffering from chronic pains; comparison between morphine, oxymorphone and anileridine. Bull World Health Organ 20:1245–1256
Eddy NB, Friebel H, Hahn KJ, Halbach H (1968) Codeine and its alternates for pain and cough relief. I. Codeine, exclusive of its antitussive action. Bull World Health Organ 38:673–741
Enggaard TP, Poulsen L, Arendt-Nielsen L, Brosen K, Ossig J, Sindrup SH (2006) The analgesic effect of tramadol after intravenous injection in healthy volunteers in relation to CYP2D6. Anesth Analg 102:146–150
Epstein DH, Preston KL, Jasinski DR (2006) Abuse liability, behavioral pharmacology, and physical-dependence potential of opioids in humans and laboratory animals: lessons from tramadol. Biol Psychiatry 73:90–99
FDA (2009) Format and content of proposed risk evaluation and mitigation strategies (REMS), REMS assessments, and proposed REMS modifications. In: U.S. Department of Health and Human Services, F.A.D.A., Center for Drug Evaluation and Research (CDER), Center for Biologics Evaluation and Research (CBER) (Ed.). U.S. Department of Health and Human Services, Rockville, MD
FDA (2010) Guidance for industry: assessment of abuse potential of drugs. In: U.S. Department of Health and Human Services, F.A.D.A., Center for Drug Evaluation and Research (CDER) (Ed.), Silver Spring, MD
FDA (2015) Abuse-deterrent opioids—evaluation and labeling guidance for industry. In: U.S. Department of Health and Human Services, F.A.D.A., Center for Drug Evaluation and Research (CDER) (ed) U.S. Department of Health and Human Services

Fischer B, Rehm J, Kirst M, Casas M, Hall W, Krausz M, Metrebian N, Reggers J, Uchtenhagen A, van den Brink W, van Ree JM (2002) Heroin-assisted treatment as a response to the public health problem of opiate dependence. Eur J Public Health 12:228–234

Fischer B, Oviedo-Joekes E, Blanken P, Haasen C, Rehm J, Schechter MT, Strang J, van den Brink W (2007) Heroin-assisted treatment (HAT) a decade later: a brief update on science and politics. J Urban Health 84:552–562

Fischer B, Jones W, Murphy Y, Ialomiteanu A, Rehm J (2015) Recent developments in prescription opioid-related dispensing and harm indicators in Ontario, Canada. Pain Physician 18:E659–E662

Fraser HF, Rosenberg DE (1964) Studies on the human addiction liability of 2′-hydroxy-5, 9-dimethyl-2-(3,3-dimethylallyl)-6, 7-benzmorphan (WIN 20, 228): a weak narcotic antagonist. J Pharmacol Exp Ther 143:149–156

Fraser HF, Van Horn GG, Martin WR, Wolbach AB, Isbell H (1961) Methods for evaluating addiction liability. (A) "Attitude of opiate addicts toward opiate-like drugs, (B) A short-term direct" addiction test. J Pharmacol Exp Ther 133:371–378

Fraser HF, Nash TL, Vanhorn GD, Isbell H (1976) Use of miotic effect in evaluating analgesic drugs in man. 1954 (proceedings). NIDA Res Monogr, p 60

Fraser HF, Kay DC, Yeh SY, Gorodetzky CW, Dewey WL (1978) Possible effects of normetabolites on the subjective and reinforcing characteristics of opioids in animals and man. Drug Alcohol Depend 3:301–318

Greenwald MK (2008) Behavioral economic analysis of drug preference using multiple choice procedure data. Drug Alcohol Depend 93:103–110

Greenwald MK, Stitzer ML (1998) Butorphanol agonist effects and acute physical dependence in opioid abusers: comparison with morphine. Drug Alcohol Depend 53:17–30

Greenwald MK, June HL, Stitzer ML, Marco AP (1996) Comparative clinical pharmacology of short-acting mu opioids in drug abusers. J Pharmacol Exp Ther 277:1228–1236

Haertzen CA (1974) An overview of addiction research center inventory scales (ARCI): an appendix and manual of scales. In: Abuse, N.I.O.D. (ed) Department of Health, Education and Welfare, Rockville, MD

Harris SC, Perrino PJ, Smith I, Shram MJ, Colucci SV, Bartlett C, Sellers EM (2014) Abuse potential, pharmacokinetics, pharmacodynamics, and safety of intranasally administered crushed oxycodone HCl abuse-deterrent controlled-release tablets in recreational opioid users. J Clin Pharmacol 54:468–477

Heishman SJ, Schuh KJ, Schuster CR, Henningfield JE, Goldberg SR (2000) Reinforcing and subjective effects of morphine in human opioid abusers: effect of dose and alternative reinforcer. Psychopharmacology 148:272–280

Hill JL, Zacny JP (2000) Comparing the subjective, psychomotor, and physiological effects of intravenous hydromorphone and morphine in healthy volunteers. Psychopharmacology 152:31–39

Horan PJ, Ho IK (1989) Comparative pharmacological and biochemical studies between butorphanol and morphine. Pharmacol Biochem Behav 34:847–854

Isbell H, Vogel VH (1949) The addiction liability of methadon (amidone, dolophine, 10820) and its use in the treatment of the morphine abstinence syndrome. Am J Psychiatry 105:909–914

Isbell H, Wikler A et al (1947) Tolerance and addiction liability of 6-dimethylamino-4-4-diphenylheptanone-3 (methadon). J Am Med Assoc 135:888–894

Jaffe JH, Martin WR (1991) Opioid analgesics and antagonists. In: Gilman AG, Rall RW, Nies AS, Taylor P (eds) Goodman and gilman's the pharmacological basis of therapeutics. Maxwell Macmillan International Editions, Elmsford, New York, pp 485–521

Jasinski DR (1977) Assessment of the abuse potentiality of morphinelike drugs (methods used in man). In: Martin WR (ed) Drug addiction I. Springer, Berlin, pp 197–258

Jasinski DR, Mansky PA (1972) Evaluation of nalbuphine for abuse potential. Clin Pharmacol Ther 13:78–90

Jasinski DR, Nutt JG (1972) Progress report on the assessment program of the NIMH addiction research center. In: U.S. Department of Health, E., and Welfare, Public Health Service, Health

Services and Mental Health Administration, National Institute of Mental Health, (ed) National, Lexington, Kentucky, pp 442–477
Jasinski DR, Preston KL (1986) Comparison of intravenously administered methadone, morphine and heroin. Drug Alcohol Depend 17:301–310
Jasinski DR, Martin WR, Hoeldtke RD (1970) Effects of short- and long-term administration of pentazocine in man. Clin Pharmacol Ther 11:385–403
Jasinski DR, Griffith JD, Pevnick JS, Clark SC (1975) Progress report on studies for he clinical pharmacology section of the addiction research center. National Academy of Sciences, Washington, DC, pp 121–161
Jasinski DR, Pevnick JS, Griffith JD (1978) Human pharmacology and abuse potential of the analgesic buprenorphine: a potential agent for treating narcotic addiction. Arch Gen Psychiatry 35:501–516
Jasinski DR, Fudala PJ, Johnson RE (1989) Sublingual versus subcutaneous buprenorphine in opiate abusers. Clin Pharmacol Ther 45:513–519
Jasinski DR, Preston KL, Sullivan JT, Testa MP (1993) Abuse potential of oral tramadol. In: 55th Annual meeting of the college on problems of drug dependence, p 103
Jones JD, Madera G, Comer SD (2014) The reinforcing and subjective effects of intravenous and intranasal buprenorphine in heroin users. Pharmacol Biochem Behav 122:299–306
Jones JD, Sullivan MA, Vosburg SK, Manubay JM, Mogali S, Metz V, Comer SD (2015) Abuse potential of intranasal buprenorphine versus buprenorphine/naloxone in buprenorphine-maintained heroin users. Addict Biol 20:784–798
Kaplan HL, Busto UE, Baylon GJ, Cheung SW, Otton SV, Somer G, Sellers EM (1997) Inhibition of cytochrome P450 2D6 metabolism of hydrocodone to hydromorphone does not importantly affect abuse liability. J Pharmacol Exp Ther 281:103–108
Kathiramalainathan K, Kaplan HL, Romach MK, Busto UE, Li NY, Sawe J, Tyndale RF, Sellers EM (2000) Inhibition of cytochrome P450 2D6 modifies codeine abuse liability. J Clin Psychopharmacol 20:435–444
Kim I, Barnes AJ, Oyler JM, Schepers R, Joseph RE Jr, Cone EJ, Lafko D, Moolchan ET, Huestis MA (2002) Plasma and oral fluid pharmacokinetics and pharmacodynamics after oral codeine administration. Clin Chem 48:1486–1496
Klous MG, Van den Brink W, Van Ree JM, Beijnen JH (2005) Development of pharmaceutical heroin preparations for medical co-prescription to opioid dependent patients. Drug Alcohol Depend 80:283–295
Lamas X, Farre M, Cami J (1994) Acute effects of pentazocine, naloxone and morphine in opioid-dependent volunteers. J Pharmacol Exp Ther 268:1485–1492
Lamb RJ, Preston KL, Schindler CW, Meisch RA, Davis F, Katz JL, Henningfield JE, Goldberg SR (1991) The reinforcing and subjective effects of morphine in post-addicts: a dose-response study. J Pharmacol Exp Ther 259:1165–1173
Lewis JW (1974) C-bridged derivatives of thebaine and oripavine. Adv Biochem Pharmacol 8:123–136
Lintzeris N, Mitchell TB, Bond AJ, Nestor L, Strang J (2007) Pharmacodynamics of diazepam co-administered with methadone or buprenorphine under high dose conditions in opioid dependent patients. Drug Alcohol Depend 91:187–194
Lofwall MR, Walsh SL (2014) A review of buprenorphine diversion and misuse: the current evidence base and experiences from around the world. J Addict Med 8:315–326
Lofwall MR, Nuzzo PA, Walsh SL (2012) Effects of cold pressor pain on the abuse liability of intranasal oxycodone in male and female prescription opioid abusers. Drug Alcohol Depend 123:229–238
Marsch LA, Bickel WK, Badger GJ, Rathmell JP, Swedberg MD, Jonzon B, Norsten-Hoog C (2001) Effects of infusion rate of intravenously administered morphine on physiological, psychomotor, and self-reported measures in humans. J Pharmacol Exp Ther 299:1056–1065
Martin WR (1983) Pharmacology of opioids. Pharmacol Rev 35:283–323
McAvoy BR, Dobbin MD, Tobin CL (2011) Over-the-counter codeine analgesic misuse and harm: characteristics of cases in Australia and New Zealand. NZ Med J 124:29–33

McNair DM, Lorr M, Droppleman LF (1971) EITS manual for the profile of mood states. Educational and Industrial Testing Service, San Diego

Mendelson J, Jones RT, Welm S, Baggot M, Fernandez I, Melby AK, Nath RP (1999) Buprenorphine and naloxone combinations: the effects of three dose ratios in morphine-stabilized, opiate-dependent volunteers. Psychopharmacology 141:37–46

Middleton LS, Nuzzo PA, Lofwall MR, Moody DE, Walsh SL (2011) The pharmacodynamic and pharmacokinetic profile of intranasal crushed buprenorphine and buprenorphine/naloxone tablets in opioid abusers. Addiction 106:1460–1473

Middleton LS, Lofwall MR, Nuzzo PA, Siegel AJ, Walsh SL (2012) Intranasal oxycodone self-administration in non-dependent opioid abusers. Exp Clin Psychopharmacol 20:310–317

Otton SV, Schadel M, Cheung SW, Kaplan HL, Busto UE, Sellers EM (1993) CYP2D6 phenotype determines the metabolic conversion of hydrocodone to hydromorphone. Clin Pharmacol Ther 54:463–472

Peters RJ Jr, Williams M, Ross MW, Atkinson J, Yacoubian GS Jr (2007) Codeine cough syrup use among African-American crack cocaine users. J Psychoactive Drugs 39:97–102

Pickworth WB, Johnson RE, Holicky BA, Cone EJ (1993) Subjective and physiologic effects of intravenous buprenorphine in humans. Clin Pharmacol Ther 53:570–576

Pircio AW, Gylys JA, Cavanagh RL, Buyniski JP, Bierwagen ME (1976) The pharmacology of butorphanol, a 3, 14-dihydroxymorphinan narcotic antagonist analgesic. Arch Int Pharmacodyn Ther 220:231–257

Preston KL, Bigelow GE (1993) Differential naltrexone antagonism of hydromorphone and pentazocine effects in human volunteers. J Pharmacol Exp Ther 264:813–823

Preston KL, Bigelow GE (1994) Drug discrimination assessment of agonist-antagonist opioids in humans: a three-choice saline-hydromorphone-butorphanol procedure. J Pharmacol Exp Ther 271:48–60

Preston KL, Walsh SL (1998) Evaluating abuse liability: methods and predictive value. In: Karch SB (ed) drug abuse handbook. CRC Press, Boca Raton, pp 276–306

Preston KL, Bigelow GE, Liebson IA (1987) Comparative evaluation of morphine, pentacozine and ciramadol in postaddicts. J Pharmacol Exp Ther 240:900–910

Preston KL, Bigelow GE, Liebson IA (1988) Butorphanol-precipitated withdrawal in opioid-dependent human volunteers. J Pharmacol Exp Ther 246:441–448

Preston KL, Bigelow GE, Bickel WK, Liebson IA (1989a) Drug discrimination in human postaddicts: agonist-antagonist opioids. J Pharmacol Exp Ther 250:184–196

Preston KL, Bigelow GE, Liebson IA (1989b) Antagonist effects of nalbuphine in opioid-dependent human volunteers. J Pharmacol Exp Ther 248:929–937

Preston KL, Bigelow GE, Liebson IA (1990) Discrimination of butorphanol and nalbuphine in opioid-dependent humans. Pharmacol Biochem Behav 37:511–522

Preston KL, Jasinski DR, Testa M (1991) Abuse potential and pharmacological comparison of tramadol and morphine. Drug Alcohol Depend 27:7–17

Preston KL, Liebson IA, Bigelow GE (1992) Discrimination of agonist-antagonist opioids in humans trained on a two-choice saline-hydromorphone discrimination. J Pharmacol Exp Ther 261

PurduePharma (2015) OXYCONTIN—oxycodone hydrochloride tablet, film coated, extended release Package Insert. In: LP, P.P. (ed)

Romach MK, Schoedel KA, Sellers EM (2014) Human abuse liability evaluation of CNS stimulant drugs. Neuropharmacology 87:81–90

Rosado J, Walsh SL, Bigelow GE, Strain EC (2007) Sublingual buprenorphine/naloxone precipitated withdrawal in subjects maintained on 100 mg of daily methadone. Drug Alcohol Depend 90:261–269

Schoedel KA, McMorn S, Chakraborty B, Potts SL, Zerbe K, Sellers EM (2011) Positive and negative subjective effects of extended-release oxymorphone versus controlled-release oxycodone in recreational opioid users. J Opioid Manag 7:179–192

Schoedel KA, Rolleri RL, Faulknor JY, Pixton GC, Chen N, Bass A, Sommerville KW, Sellers E (2012) Assessing subjective and physiologic effects following intranasal administration of a

new formulation of immediate release oxycodone HCl (Oxecta) tablets in nondependent recreational opioid users. J Opioid Manag 8:315–327
Schuh KJ, Walsh SL, Bigelow GE, Preston KL, Stitzer ML (1996) Buprenorphine, morphine and naloxone effects during ascending morphine maintenance in humans. J Pharmacol Exp Ther 278:836–846
Soergel DG, Subach RA, Burnham N, Lark MW, James IE, Sadler BM, Skobieranda F, Violin JD, Webster LR (2014) Biased agonism of the mu-opioid receptor by TRV130 increases analgesia and reduces on-target adverse effects versus morphine: a randomized, double-blind, placebo-controlled, crossover study in healthy volunteers. Pain 155:1829–1835
Stoops WW, Hatton KW, Lofwall MR, Nuzzo PA, Walsh SL (2010) Intravenous oxycodone, hydrocodone, and morphine in recreational opioid users: abuse potential and relative potencies. Psychopharmacology 212:193–203
Stoops WW, Lofwall MR, Nuzzo PA, Craig LB, Siegel AJ, Walsh SL (2012) Pharmacodynamic profile of tramadol in humans: influence of naltrexone pretreatment. Psychopharmacology 223:427–438
Stoops WW, Glaser PE, Rush CR (2013) Miotic and subject-rated effects of therapeutic doses of tapentadol, tramadol, and hydromorphone in occasional opioid users. Psychopharmacology 228:255–262
Strain EC, Preston KL, Liebson IA, Bigelow GE (1992) Acute effects of buprenorphine, hydromorphone and naloxone in methadone-maintained volunteers. J Pharmacol Exp Ther 261:985–993
Strain EC, Preston KL, Liebson IA, Bigelow GE (1993) Precipitated withdrawal by pentazocine in methadone-maintained volunteers. J Pharmacol Exp Ther 267:624–634
Strain EC, Preston KL, Liebson IA, Bigelow GE (1995) Buprenorphine effects in methadone-maintained volunteers: effects at two hours after methadone. J Pharmacol Exp Ther 272:628–638
Strain EC, Walsh SL, Preston KL, Liebson IA, Bigelow GE (1997) The effects of buprenorphine in buprenorphine-maintained volunteers. Psychopharmacology 129:329–338
Strain EC, Stoller K, Walsh SL, Bigelow GE (2000) Effects of buprenorphine versus buprenorphine/naloxone tablets in non-dependent opioid abusers. Psychopharmacology 148.374–383
Sullivan MA, Vosburg SK, Comer SD (2006) Depot naltrexone: antagonism of the reinforcing, subjective, and physiological effects of heroin. Psychopharmacology 189:37–46
Toll L, Bruchas MR, Calo G, Cox BM, Zaveri NT (2016) Nociceptin/orphanin FQ receptor structure, signaling, ligands, functions, and interactions with opioid systems. Pharmacol Rev 68:419–457
Tompkins DA, Smith MT, Bigelow GE, Moaddel R, Venkata SL, Strain EC (2014) The effect of repeated intramuscular alfentanil injections on experimental pain and abuse liability indices in healthy males. Clin J Pain 30:36–45
Umbricht A, Huestis MA, Cone EJ, Preston KL (2004) Effects of high-dose intravenous buprenorphine in experienced opioid abusers. J Clin Psychopharmacol 24:479–487
Van Zee A (2009) The promotion and marketing of oxycontin: commercial triumph, public health tragedy. Am J Public Health 99:221–227
Volpe DA, McMahon Tobin GA, Mellon RD, Katki AG, Parker RJ, Colatsky T, Kropp TJ, Verbois SL (2011) Uniform assessment and ranking of opioid mu receptor binding constants for selected opioid drugs. Regul Toxicol Pharmacol 59:385–390
Walker DJ, Zacny JP (1998) Subjective, psychomotor, and analgesic effects of oral codeine and morphine in healthy volunteers. Psychopharmacology 140:191–201
Walker DJ, Zacny JP (1999) Subjective, psychomotor, and physiological effects of cumulative doses of opioid mu agonists in healthy volunteers. J Pharmacol Exp Ther 289:1454–1464
Walsh SL, Preston KL, Stitzer ML, Cone EJ, Bigelow GE (1994) Clinical pharmacology of buprenorphine: ceiling effects at high doses. Clin Pharmacol Ther 55:569–580

Walsh SL, June HL, Schuh KJ, Preston KL, Bigelow GE, Stitzer ML (1995a) Effects of buprenorphine and methadone in methadone-maintained subjects. Psychopharmacology 119:268–276
Walsh SL, Preston KL, Bigelow GE, Stitzer ML (1995b) Acute administration of buprenorphine in humans: partial agonist and blockade effects. J Pharmacol Exp Ther 274:361–372
Walsh SL, Strain EC, Abreu ME, Bigelow GE (2001) Enadoline, a selective kappa opioid agonist: comparison with butorphanol and hydromorphone in humans. Psychopharmacology 157: 151–162
Walsh SL, Chausmer AE, Strain EC, Bigelow GE (2008a) Evaluation of the mu and kappa opioid actions of butorphanol in humans through differential naltrexone blockade. Psychopharmacology 196:143–155
Walsh SL, Nuzzo PA, Lofwall MR, Holtman JR Jr (2008b) The relative abuse liability of oral oxycodone, hydrocodone and hydromorphone assessed in prescription opioid abusers. Drug Alcohol Depend 98:191–202
Walsh SL, Heilig M, Nuzzo PA, Henderson P, Lofwall MR (2012) Effects of the NK1 antagonist, aprepitant, on response to oral and intranasal oxycodone in prescription opioid abusers. Addict Biol e-pub ahead of print
Webster LR, Bath B, Medve RA, Marmon T, Stoddard GJ (2012) Randomized, double-blind, placebo-controlled study of the abuse potential of different formulations of oral oxycodone. Pain Med 13:790–801
Weinhold LL, Preston KL, Farre M, Liebson IA, Bigelow GE (1992) Buprenorphine alone and in combination with naloxone in non-dependent humans. Drug Alcohol Depend 30:263–274
White KL, Robinson JE, Zhu H, DiBerto JF, Polepally PR, Zjawiony JK, Nichols DE, Malanga CJ, Roth BL (2015) The G protein-biased kappa-opioid receptor agonist RB-64 is analgesic with a unique spectrum of activities in vivo. J Pharmacol Exp Ther 352:98–109
Wikler A, Fraser HF, Isbell H (1953) N-Allylnormorphine: effects of single doses and precipitation of acute abstinence syndromes during addiction to morphine; methadone or heroin in man (post addicts). J Pharmacol Exp Ther 109:8–20
Zacny JP, Lichtor SA (2008) Within-subject comparison of the psychopharmacological profiles of oral oxycodone and oral morphine in non-drug-abusing volunteers. Psychopharmacology 196:105–116
Zacny JP, Gutierrez S (2009) Within-subject comparison of the psychopharmacological profiles of oral hydrocodone and oxycodone combination products in non-drug-abusing volunteers. Drug Alcohol Depend 101:107–114
Zacny JP, Lichtor JL, Zaragoza JG, de Wit H (1992) Subjective and behavioral responses to intravenous fentanyl in healthy volunteers. Psychopharmacology 107:319–326
Zacny JP, Lichtor JL, Binstock W, Coalson DW, Cutter T, Flemming DC, Glosten B (1993) Subjective, behavioral and physiological responses to intravenous meperidine in healthy volunteers. Psychopharmacology 111:306–314
Zacny JP, Lichtor JL, Thapar P, Coalson DW, Flemming D, Thompson WK (1994) Comparing the subjective, psychomotor and physiological effects of intravenous butorphanol and morphine in healthy volunteers. J Pharmacol Exp Ther 270:579–588
Zacny JP, McKay MA, Toledano AY, Marks S, Young CJ, Klock PA, Apfelbaum JL (1996) The effects of a cold-water immersion stressor on the reinforcing and subjective effects of fentanyl in healthy volunteers. Drug Alcohol Depend 42:133–142
Zacny JP, Gutierrez S, Bolbolan SA (2005) Profiling the subjective, psychomotor, and physiological effects of a hydrocodone/acetaminophen product in recreational drug users. Drug Alcohol Depend 78:243–252

Over-the-Counter Codeine—from Therapeutic Use to Dependence, and the Grey Areas in Between

Suzanne Nielsen and Marie Claire Van Hout

Abstract Codeine is a widely used analgesic, that is available for sale in pharmacies over the counter (OTC) in a number of countries including the UK, South Africa, Ireland, France and Australia. In these countries with OTC codeine sales there has been emerging concerns about misuse of and dependence on codeine containing combination analgesics, with increasing numbers of people presenting for help with codeine dependence at primary care and addiction treatment services. This has led to many countries reviewing availability of codeine in OTC available preparations, and considering possible measures to reduce harms from misuse of OTC codeine containing combination analgesics.

Keywords Codeine · Dependence · Opioid use disorder · Analgesic · Abuse

Contents

S. Nielsen (✉)
National Drug and Alcohol Research Centre, UNSW, 22-32 King Street, Randwick, NSW 2031, Australia
e-mail: suzanne.nielsen@unsw.edu.au

S. Nielsen
South East Sydney Local Health District (SESLHD) Drug and Alcohol Services, 591-623 S Dowling Street, Surry Hills, NSW 2010, Australia

M.C. Van Hout
Department of Health, Sport and Exercise Science, School of Health Sciences, Waterford Institute of Technology, Waterford, Ireland

Curr Topics Behav Neurosci (2017) 34:59–75
DOI 10.1007/7854_2015_422
Published Online: 15 January 2016

1 Codeine Pharmacology

Codeine is the most commonly used opioid analgesic in the world (International Narcotics Control Board 2012) and has long been used for relief of mild-to-moderate pain, as an antitussive and antidiarrheal. Codeine is a methylated morphine derivative that is found naturally along with morphine in the poppy seed. Codeine is most commonly taken orally and has 30–40 % oral bioavailability (Leppert 2011). Once absorbed, the o-demethylation of codeine into morphine occurs (primarily via CYP2D6 enzymes) into morphine, and this process is thought to be largely responsible for its analgesic activities (Dayer et al. 1988), though other metabolites also display analgesic activity (Leppert 2011). Codeine displays common opioid-related side effects, including constipation, sedation and euphoria.

The amount and rate at which morphine is converted varies between individuals. Approximately 10 % of the Caucasian population are thought to be poor or slow metabolisers of codeine and are unable to metabolise codeine sufficiently to elicit an analgesic response. Around 1–7 % of Caucasians are thought to be ultrafast metabolisers who convert codeine at a higher than average rate due to the presence of multiple and active CYP2D6 genes resulting in potential opioid toxicity (Madadi et al. 2013). The presence of ultrafast and slow metabolism varies in the general population, notably between different ethnic groups (Zhou 2009).

2 Codeine Availability

Globally, there is wide variation in the regulation of codeine availability. In many countries, codeine is only available through prescription. In contrast, in countries such as Australia, New Zealand, Canada, South Africa, France, Ireland and the United Kingdom codeine is widely available over-the-counter (OTC, i.e. for sale without a prescription), though there are differing restrictions around advertising, shelf-visibility and level of pharmacist intervention required for sales. Codeine is usually available in combination products (e.g. compounded with paracetamol or ibuprofen when sold over-the-counter). Conversely, France has codeine available as a single-ingredient over-the-counter product, removing the concerns associated with consuming high doses of combination products. Although systematic data on non-medical codeine use are not available to enable intercountry comparisons, reports of codeine-related harms have largely been published from these countries with fewer restrictions on codeine availability, and centre on the harms relating of additives such as paracetamol and ibuprofen. For example, serious harms have been reported from consumption of high doses of ibuprofen–codeine combinations in Australia (Dutch 2008; Frei et al. 2010) and the United Kingdom (Dyer et al. 2004; Medani et al. 2010), New Zealand (Robinson et al. 2010; Evans et al. 2010; McAvoy et al. 2011), and increasing substance abuse treatment demands have been attributed to OTC codeine (Myers et al. 2003; Nielsen et al. 2015, 2014).

3 Non-medical Use and Dependence

The spectrum of non-medical-use of pharmaceuticals includes problematic consumption outside of acceptable medical practice or medical guidelines, when self-medicating at higher doses and for longer than advised, for recreational intoxicating purposes and when risk and adverse consequences outweigh benefits (Casati et al. 2012). Efforts to quantify the extent and understand this form of non-medical or problematic pharmaceutical drug as distinct from illicit drug use, particularly in relation to codeine, are complicated by retail availability, lack of monitoring and surveillance, and the hidden and heterogeneous nature of therapeutic and non-therapeutic forms of misuse (Casati et al. 2012; Cooper 2013a, 2011; Lessenger and Feinberg 2008). Factors underpinning non-medical use of pharmaceuticals more generally centre on a myriad of individual, environmental and geographic factors relating to availability, inappropriate and excessive prescribing, misconceptions around safety, self-medication of physical and emotional pain, recreational use, accessing of multiple prescribers and pharmacies, illicit and Internet sourcing, scheduling and pharmaceutical marketing ploys (Maxwell 2011; Boyd and McCabe 2008; McCabe and Boyd 2005; Nordmann et al. 2013).

Codeine's identified abuse potential has been demonstrated in drug administration studies (Babalonis et al. 2013) and also by reported cases of dependence (Frei et al. 2010; Sproule et al. 1999). Although abuse liability may be lower than many prescription opioid with stronger opioid effects (Babalonis et al. 2013), consistent with other opioids, neuro-adaptation to its opioid effects occurs with repeated administration within a relatively short period of time (Nielsen et al. 2010; Toms et al. 2009). Increasing doses for therapeutic, non-medical or recreational purposes increases the likelihood of physiological neuro-adaptation and the development of dependence (Nielsen et al. 2010; Reed et al. 2011). Physical tolerance over time can lead to negative physical effects on cessation which include, insomnia, restlessness, runny nose, stomach pains, diarrhoea and chills. This is in addition to psychological symptoms of opioid dependence which may include cravings, preoccupation with securing and consuming codeine and a lack of control over use despite experiencing harms (Nielsen et al. 2010; Vallejo and Wang 2011). Public availability of non-prescription codeine-containing preparations along with the encouragement of consumer self-treatment, and the lack of awareness of habit-forming use and related harms, has contributed to the potential for diverse patterns of non-medical use and dependence (Wazaify et al. 2005; Van Hout et al. 2014; Cooper 2013b). In terms of distinguishing different patterns of codeine use, two independent studies conducted in Australia and the United Kingdom identified three broad categories of problematic codeine use: (1) use which never exceeds the maximum recommended dose, though in terms of duration and nature of use meets criteria for dependence, (2) consumption of slightly higher than the recommended dose (either in a therapeutic or recreational context) and (3) consumption doses that substantially exceed recommended doses (largely in the context of serious opioid dependence) (Nielsen et al. 2010; Cooper 2011). Many of these consumption patterns can fall into the grey area between therapeutic use and non-medical use, with consumers themselves having difficulties identifying problematic use along their own trajectory of codeine use, as demonstrated in the example below (See Box 1).

Box 1: Case example Janette was a 32-year-old personal assistant, working in a new job with a demanding employer. She began suffering regular headaches and started taking an ibuprofen–codeine product for relief. Initially she was taking the ibuprofen–codeine for pain, but after she began taking it she also realised that she felt more relaxed after a dose. Slowly she began taking it more often, for headaches and period pain, until she was taking it every couple of days, and then every day. Initially she was just taking them two at a time, three times a day, never exceeding the maximum dose. After a few months, her job became more stressful and her headaches seemed to worsen. She increased to dose to six tablets at a time, and then six tablets twice a day, rationalising if you could take six tablets in a day safely, then it must be ok to take them all at once.

Before she knew it, she was taking more than the recommended daily dose but 'just turning a blind eye'. She found it hard to get through a day without codeine, and the dose kept slowly creeping up when she had a stressful day or more headaches. She attended to multiple pharmacies who started to ask more questions in relation to why she needed the codeine, but she always answered that she was taking it for headaches and that her headaches always improved after taking codeine, and she was never denied a sale. As time progressed, she became more and unwell, but felt embarrassed to let her friends and family know how much medication she was taking and justified to herself that if it was not safe, they would not sell it in the pharmacy.

Eventually Janette presented to the emergency department after an episode of vomiting with blood in it. She was admitted with a duodenal haemorrhage and anaemia. The emergency department doctor had previously seen several similar cases and questioned her about her non-prescribed medication use. After disclosing her current use (now 48 tablets a day), the doctor referred her to speak to the drug and alcohol team, who were able to explain about the symptoms of opioid dependence and talked through some treatment options with her.

Misuse of codeine-containing preparations, and resultant codeine dependence, can occur for both legitimate (therapeutic) and intoxicating (non-therapeutic) purposes (Frei et al. 2010; Nielsen et al. 2014; Van Hout 2015; McDonough 2011). Studies underscore the interplay between self-medication, chronic pain and iatrogenic dependence in the case of misuse of codeine-containing products (Nielsen et al. 2014; Hamer et al. 2014; Roussin et al. 2013). Individuals dependent on codeine are more likely to report chronic pain, and more likely to use codeine to reduce stress and for its pleasurable and relaxing effect (Sproule et al. 1999). Nielsen et al. (2010) described how individuals with '*iatrogenic*' dependence following medical use of codeine for pain, anxiety or insomnia may experience a '*blurring*' between therapeutic and problematic use over time, identifying codeine dependent people who were aware of their dependence but continuing to use in response to cravings and in avoiding withdrawal symptoms in addition to those unknowingly misusing codeine falsely thinking that they were medicating their physical condition where in fact they were medicating opioid withdrawal symptoms.

Reports of non-medical use and dependence are observed to occur in a diverse range of populations. These include parental medication of children with codeine-containing products (Allotey et al. 2004), codeine misuse in psychiatric patients (Agyapong et al. 2013) and in populations of substance users (Armstrong 1992; Arora et al. 2013). Recreational use of codeine cough mixtures for intoxicating purposes is reported among youth and drug users (e.g. '*purple drank*') (Van Hout et al. 2014; Kitabayashi et al. 2000; Roumie 2004; Otto et al. 2009; MacFadyen and McGowan 2001) and Internet drug forum users (Van Hout 2015). Case reports also exist where misuse of codeine was identified as an iatrogenic cause of psychiatric disturbances (Manchia et al. 2013).

Despite the aforementioned studies identifying diverse groups people who report non-medical codeine use, and individuals dependent on codeine, the hidden and heterogeneous nature of non-medical codeine use remains a complicating factor, and current limited estimations of prevalence rely on treatment uptake and database studies (Roussin et al. 2013; Pates et al. 2002; Skurtveit et al. 2011). Being able to differentiate between requests for codeine-containing products which represent safe and legitimate therapeutic use from those where problematic use may be developing confounds recognition of individuals experiencing problems. This is further complicated by those who are developing problems not readily identifying themselves as needing help or as a '*drug addict*' (Cooper 2013a, b; Nielsen et al. 2010; Pates et al. 2002). Cooper (2013a) reported that people who use codeine, despite recognising they were dependent on codeine, reported regret on loss of control of their codeine use and associated work and health problems, but self-identified as different to people who use illicit drugs by virtue of continued social and economic activity. Individuals dependent on codeine in Australia have indeed been found to differ from other populations of individuals dependent on prescription opioids (oxycodone, morphine) and heroin by higher levels of employment and being more likely to be female (Nielsen et al. 2014, 2011). A comparison of people who recently used over-the-counter codeine found meeting criteria for dependence was associated with a younger age, lower levels of employment and education, mental health problems and with family history of problematic substance use (Nielsen et al. 2011). Those seeking treatment for codeine in Australia have also been prominently older females, distinguishing them from traditional populations of opioid-dependent people, though this pattern has been changing over time with an increase in younger males presenting from treatment (Nielsen et al. 2015).

Further research is warranted to identify risk profiles of developing problematic codeine use, to garner greater understanding of therapeutic and non-therapeutic pathways and trajectories to misuse and dependence. This work may inform more effective strategies for prevention and to understand how consumer displacement between legitimate pharmacy supply and illicit sourcing via diversion or web retail occurs (Casati et al. 2012). One of the challenges is that with the diversity of presentations of codeine dependence and problematic use, targeting interventions to specific populations is quite difficult, suggesting broader population-level responses are required.

4 Harms from High-Dose Consumption of Over-the-Counter Codeine

Clinical case studies have reported on adverse health consequences relating to excessive or long-term misuse of combination codeine analgesics containing additives (ibuprofen, paracetamol and aspirin). Chronic headache, gastrointestinal haemorrhage, nephrotoxicity, hypokalaemia, acute haemorrhagic necrotising pancreatitis, neurological damage and addiction are reported, and of note often in individuals with no history of substance use disorders and comorbidity (Dutch 2008; Frei et al. 2010; McAvoy et al. 2011; McDonough 2011; Chetty et al. 2003; Ernest et al. 2010; Pilgrim et al. 2014; Ng et al. 2011). A summary of some of the most common harms is outlined in Table 1.

4.1 Evidence for Effectiveness

Continued debate around availability of non-prescription codeine in particular centres not only on abuse potential, but also on adverse health effects associated with the presence of simple analgesics (paracetamol, ibuprofen) in combination with codeine, and the lack of pharmacological evidence to support effectiveness of low- dose codeine in combination analgesics (Ferner and Beard 2008; Nielsen et al. 2012; Murnion 2010). Standards advise that products containing codeine should only be supplied when single-ingredient medicinal products are ineffective, as '*second-line*' products for the treatment of pain and only used in accordance with marketing authorisations for short-term use (no longer than three days). The World

Table 1 Common harms associated with over-the-counter codeine use

Harm	Description	Main products	Published cases
Opioid dependence	Features of dependence include escalating doses, withdrawal symptoms on discontinuation, and continued use of codeine despite experiencing harms. A number of cases of dependence to OTC codeine products have been successfully treated with opioid agonist treatments such as methadone or buprenorphine	OTC codeine combination products	Frei et al. (2010), McDonough (2011), Robinson et al. (2010), Evans et al. (2010)
Gastrointestinal harm	Gastric and peptic ulcers, perforation, haemorrhage, pyloric stenosis, gastrectomy and other bowel surgery	OTC codeine–ibuprofen	Dutch (2008), Frei et al. (2010), McDonough (2011), Evans et al. (2010), Robinson et al. (2010)
Anaemia	Resulting from blood loss, often secondary to gastric ulcers	OTC codeine–ibuprofen	Frei et al. (2010), Evans et al. (2010), Robinson et al. (2010)
Renal effects	Renal tubular acidosis due to prolonged high doses of ibuprofen, renal failure	OTC codeine–ibuprofen	Frei et al. (2010)
Hypokalaemia	Low potassium leading to muscle weakness and respiratory difficulties, one case of rhabdomyolysis and quadriparesis. Some cases requiring ICU admission for life support	OTC codeine–ibuprofen	Ernest et al. (2010), Frei et al. (2010), Ng et al. (2011)

Health Organisation has placed codeine as a '*step 2*' on its pain ladder. Recent debates centre on the suggestion to skip '*step 2*' due to problems with codeine (and tramadol), due to limited evidence of effectiveness, variations in patient metabolism and availability of more predictable opioids (Vargas-Schaffer 2010). Cochrane reviews have underscored the lack of data to support low-dose codeine (<10 mg) and limited data to support medium-dose (10–20 mg) codeine for analgesic efficiency, with combined ibuprofen (400 mg) and codeine (25.6–60 mg) incurring good analgesic efficiency (Toms et al. 2009). There is also limited evidence for single-dose oral ibuprofen plus codeine being more effective for post-operative pain than either drug in isolation (Baratta et al. 2014). A meta-analysis of opioids for osteoarthritis of the knee or hip reported that modest benefits of codeine were outweighed by adverse consequences (Nuesch et al. 2009). Given the low dose of codeine in non-prescription medicines (equivalent often to approximately 2 mg of morphine), non-opioid analgesics may often perform better (Murnion 2010). The evidence for efficacy must therefore be considered when trying to create the right balance between availability and harms.

5 Responses to Problematic Over-the-Counter Codeine Use

5.1 Increased Restrictions on Sales

Misuse of pharmaceuticals is a public health issue amid increasing calls for (up) scheduling and enhanced pharmacovigilance in clinical and pharmacy practice initiatives (Casati et al. 2012; UVUN 2011). Global awareness around misuse of non-prescription codeine-containing products is increasing (McAvoy et al. 2011; Myers et al. 2003; Roussin et al. 2013; Orriols et al. 2009). Given the diverse characteristics of people who use codeine in non-medical ways, along with the continuum of use of non-medical codeine ranging from use for reasons other than pain, recreational use and iatrogenic dependence on codeine, there is a public health imperative to develop effective responses. Strategies to ensuring dispensing, risk screening, management, monitoring and surveillance of prescribed and non-prescription codeine use. Consideration of alternative pharmacovigilant approaches to enhance patient information and reduce misuse and dependence on non-prescription codeine-containing products are warranted.

The debate around codeine centres on the availability of non-prescription products containing codeine to the general public (Robinson et al. 2010; Cooper 2013a). Deregulation of codeine to pharmacy over-the-counter status in certain countries evolved in order to encourage patient self-treatment and reduce governmental drug budgets and general practitioner workload (Cooper 2011; Hamer et al. 2014; MacFadyen and McGowan 2001; Wazaify et al. 2006; Hughes et al. 1999; Fleming et al. 2004; McBride et al. 2003). Currently, regulatory responses have largely focused on upscheduling and improved guidelines for restricted supply of non-prescription codeine-containing preparations (Cooper 2013a; Hamer et al. 2014) and have met with resistance from the pharmaceutical industry (McAvoy et al. 2011). Upscheduling has contributed to increased visibility of warnings on labels and leaflets (i.e. '*Can cause addiction. For 3 days use only*' and '*Risk of addiction. Do not use for more than 3 days unless on medical advice*') (Medicines and Health Care Products Regulatory Agency 2009), restrictions on pack sizes, advertising codes, prevention of public advertising, restrictions on product visibility and customer self-selection, pharmacy record keeping and direct involvement of pharmacists in sales (Cooper 2013a, 2011; Hamer et al. 2014). Other mitigating actions include sale of smaller pack sizes and dispensing of lower dosages of medications containing codeine (Bateman 2013). Debate around merits of upscheduling and restricted protocols for sale continue are in response to a desire to curtail non-medical use and inappropriate use of codeine. The potential impact on restriction of supply may include reduced consumer choice for pain medication, possible under treatment of pain, potential for displacement towards stronger opioids, repeat accessing of multiple pharmacies and overburdening of healthcare systems (Le Roux 2013; LA Gudin 2013). It is vital that scheduling and related ethical and scientific factors in access and convenience are guided by safe medical decision-making and evidence-based treatment of pain (Schatman and Darnall 2013). Advocates of upscheduling codeine products highlight the potential preventative effects in avoiding opioid dependence, and reducing inappropriate patient self-medication which has been linked to serious harms including mortality involving combination codeine products (Pilgrim et al. 2014; Kolodny 2013; McAvoy and Tobin 2014). In the case of over-the-counter codeine, those supporting upscheduling, with recommendations that codeine be available only on prescription, focus on the serious harm in comparison with the limited evidence of efficacy of doses in over-the-counter codeine products (Pilgrim et al. 2014; McAvoy and Tobin 2014).

5.2 *Harm Reduction*

The role of harm reduction in this public health issue remains largely unexplored. Strategies such as selling codeine as a single ingredient have the potential to reduce harms associated with consumption of high-dose simple analgesics (paracetamol, aspirin, ibuprofen) in the context of codeine dependence. Use practices such as cold water extraction, which allows separation of paracetamol and codeine ingredients, may reduce harms for those intentionally using codeine in a non-medical way, and have been documented to be successfully used during recreational codeine use (Nielsen et al. 2010; Van Hout 2015). How to effectively disseminate this information to the broad demographic of people consuming high doses of combination codeine products may be more of a challenge, as many are not in contact with substance using peers, or in contact with harm reduction services which traditionally target people who use illicit substances. Although some people who use codeine seek information through Internet forums with the intention of reducing harms, other people who use codeine may not identify their use as high risk (Nielsen et al. 2010; Van Hout 2015).

5.3 *Challenges for Pharmacists*

Pharmacists are a trusted source of information and advice around over-the-counter medications and strongly influence public perceptions of safety and purchase behaviours in pharmacy retail (Wazaify et al. 2005; French 2008; Wawruch et al. 2013). Conflict, however, exists between retail, medical and pharmaceutical sources of information and complicate the consumer decision-making process (Hughes et al. 1999). Pharmacist reporting of concern around customer misuse and dependence on codeine-containing products (with mixed views on safety) in recent years are evident (Cooper 2011, 2013b; Wazaify et al. 2005; Roumie 2004; Pates et al. 2002; Ferner and Beard 2008; Matheson et al. 2002). In concert with efforts around scheduling, surveillance and restricted sales, the provision of codeine-containing products with appropriate patient information to reduce risk and harm is desirable (Cooper 2013b; Roumie 2004). Part of the pharmacists responsibility in codeine supply is customer screening and advice, identification of the condition, and confirming the therapeutic need for a codeine-containing product and suitability of the preparation and amount dispensed (Hamer et al. 2014; Le Roux 2013). Suspected problematic use and dependence in pharmacies often centres on customer appearance, observed high frequency of repeated requests, requesting certain codeine-containing products by name, requesting larger pack sizes, refusal to consider single-ingredient products (i.e. paracetamol, aspirin or ibuprofen), and agitation when pharmacists intervene (Hamer et al. 2014; Nielsen et al. 2013). Pharmacy responses in the event of suspected misuse include medicines information provision by medicines counter assistants (MCA) and pharmacists, direct pharmacist intervention by additional customer questioning, removal of codeine-containing products displayed at point of sale, refusal of sale or restriction of quantity sold in the event of suspect requests, on-site recording of incidences of suspected misuse and customer referral to primary care professionals (Nielsen et al. 2010; Hamer et al. 2014; MacFadyen and McGowan 2001; Pates et al. 2002). However, even in circumstances of frequent observation of customer opioid intoxication and aberrant opioid behaviours in their pharmacies, pharmacists report difficulties in the communication of their concerns to doctors (Kahan et al. 2011). Also despite such guidelines for restricted sales, relative ease of purchasing is reported by individuals misusing codeine-containing products, with refusal at one pharmacy contributing to patterns of '*pharmacy hopping*' (Van Hout 2015; Nielsen et al. 2013). These studies also described customer tactics in securing a successful purchase as centring on appropriate appearance and prerehearsed scripts in the event of pharmacist interrogation. Of note is that simple refusal of sale could be viewed as contra to duty of care, particularly when under treatment of chronic pain or opiate withdrawal is experienced by the patient (Hamer et al. 2014). Positive outcomes resulting from regulatory restrictions on supply, reduced visibility of products and direct pharmacist interventions include reduced non-medical use (Agyapong et al. 2013) and patient reports that pharmacists intervention is a factor in treatment seeking (Nielsen et al. 2013).

Challenges have been raised with pharmacy supply such as difficulties raising sensitive topics, and reluctance by codeine-dependent people to engage in such discussion, possibly because they do not self-identify or recognise their use as problematic at the time. Research has found that pharmacists are not sure about when and how to intervene, and describing a lack of information about customers to support decisions about supply (Cooper 2011). Substance use and mental health are areas that are identified as challenging: A study examining pharmacists' attitudes towards discussing psychiatric medications with patients, compared with cardiovascular medications, found that pharmacists were less comfortable discussing medications and symptoms for mental health conditions, with issues of adequate privacy, lack of training identified as barriers (Phokeo et al. 2004). Butler and Sheriden (2010a) also identified that a lack of training may contribute to pharmacists' reluctance to deliver counselling and harm reduction intervention on these sensitive topics.

5.4 *Screening and Brief Intervention*

Pharmacist intervention in the event of suspected customer misuse is associated with increased help-seeking behaviour (Nielsen et al. 2013). To date, pharmacy practice efforts have piloted harm minimisation models within integrated collaborative approaches using targeted patient identification and recruitment, treatment/referrals and data measurement in tackling OTC misuse (Wazaify et al. 2006; Fleming et al. 2004), though these approaches are limited by customer ability to access multiple pharmacies, and the infrequent detection of problematic use (Hamer et al. 2014; Nielsen et al. 2012, 2013). In this way, universal precautions approaches, which have been developed for use in responding to prescription opioid dependence, may be appropriate for expansion to community pharmacy settings to enhance reach and patient awareness, where all customers are screened for misuse (not just those suspected) (Gourlay et al. 2005). Universal precautions approaches also address issues of the timing of pharmacist intervention in codeine sales. Research with codeine-dependent people identified that those pharmacists only intervene once patterns of frequent purchase have been established (Nielsen et al. 2013), missing a valuable opportunity for early intervention or prevention. A 'universal precautions' approach, where all patients are screened for risk, and a minimal standard of precautions is applied to all patients has been suggested for prescription medications and may be able to be adapted for use with OTC codeine (Gourlay et al. 2005). This approach has the potential to reduce stigma and ensure that a minimal level of protection is applied to all patients. Further, the frequency questioning about use of codeine products has been reported by codeine-dependent people to be one of the reasons that they eventually sought help for their codeine use, highlighting the systematic questioning for all codeine sales (Nielsen et al. 2013).

The evidence base regarding delivery of patient information and healthcare advice, screening and brief intervention relating to smoking, hazardous/harmful problem drinking in primary care and pharmacy settings is compelling (Dhital et al. 2010; Zahradnik et al. 2009; Wilk et al. 1997; Strobbe 2014; Sheridan et al. 2008; O'Donnell et al. 2013; Kypri et al. 2008; Heather et al. 2004) and warrants expansion to explore it effectiveness in responding to concerns with non-medical or problematic use of over-the-counter codeine-containing preparations (Van Hout et al. 2014).

Further work on developing tailored initiatives for people who use codeine that can demonstrate enduring effects is needed. For example, brief interventions for people who use pharmaceutical opioids delivered in general hospitals show promising short-term (3 month) outcomes, although benefits are not retained at one-year follow-up (Otto et al. 2009; Zahradnik et al. 2009). Continued efforts to expand on pharmacist and pharmacy assistant training in addiction, mental health, communication skills, screening and brief intervention skills (Dhital et al. 2010; Sheridan et al. 2008; Horsfield et al. 2011; Khan et al. 2013) alongside regular training and process auditing according to regulatory protocols for supply warrant consideration (Hamer et al. 2014; Butler and Sheridan 2010b). Building on screening and brief intervention further integrated community pharmacy and primary care treatment approaches for individuals experiencing codeine dependence could be explored (Van Hout et al. 2014; Hughes et al. 1999). Buprenorphine detoxification as well as maintenance of buprenorphine and methadone has been used for the treatment of codeine dependence (Frei et al. 2010; Nielsen et al. 2014), though research to understand longer-term treatment outcomes for people who have developed dependence to codeine is warranted, particularly as they often appear to receive shorter-term treatments such as detoxification, which as not as well supported by treatment research (Nielsen et al. 2014).

5.5 Monitoring of Sales as Part of a Comprehensive Response

A multifaceted approach by all stakeholders to include pharmaceutical manufacturers, wholesalers, prescribers, pharmacists, treatment providers, law enforcement and drug education specialists is required to avoid penalising patients who benefit from prescribed pharmaceuticals such as codeine (Wilsey and Prasad 2012). Development of national online prescription systems can curtail patterns of pharmaceutical misuse (Maxwell 2011). National integrated prescriber and pharmacy monitoring of medicines using innovative real-time reporting (RTR) analysis has emerged in the USA, Canada, South Africa and Australia (UVUN 2011; Le Roux 2013; Shand et al. 2013). Well-designed RTR systems aim to monitor and track levels of dispensing, counter inappropriate prescribing, prescription fraud, shopping and unsanctioned use, curtail pharmacy hopping and ultimately incidence of adverse events such as overdose (Bateman 2013; Le Roux 2013; Wrobel 2003; Chee and Schneberger 2003). However, these systems for the most part have not yet incorporated codeine-containing preparations and require extensive evaluation and process monitoring to avoid consumer displacement towards stronger opioids, alcohol, other psychotropic drugs and the shift towards overly cautious prescribing of higher schedule opioids and difficulties in doctor responses to real-time information (Shand et al. 2013). That said, the live reporting of codeine customer consumption history at pharmacy point of sale in RTRs along with enhanced customer awareness of monthly thresholds for codeine dispensing presents a timely (and legitimate) opportunity for the pharmacist to question the sale, engage the customer in brief screening and intervention and provide referral options if required. Innovations in South Africa (see '*Codeine Care Initiative*'[1]) have taken this opportunity one step further by developing smartphone apps to provide customers with discrete access to screening for problematic use, personal purchasing history and product-related information. RTR systems expanded to include codeine also offer public health policy makers and practitioners a unique mechanism to record codeine consumption trends, and so monitor for misuse and dependence.

Funding Acknowledgement Suzanne Nielsen is supported by NHMRC fellowship (#1013803). The National Drug and Alcohol Research Centre at the University of NSW is supported by funding from the Australian Government under the Substance Misuse Prevention and Service Improvements Grant Fund. Research leading to these results has received funding from the European Community's Seventh Framework Programme FP7/2007–2013 under Grant Agreement No 611736.

References

Agyapong VIO, Singh K, Savage M, Thekiso TB, Finn M, Farren CK et al (2013) Use of codeine-containing medicines by Irish psychiatric inpatients before and after regulatory limitations on their supply. Ir J Psychol Med 30(01):7–12

Allotey P, Reidpath DD, Elisha D (2004) "Social medication" and the control of children: a qualitative study of over-the-counter medication among Australian children. Pediatrics 114(3): e378–e383

Armstrong DJ (1992) The use of over-the-counter preparations by drug users attending an addiction treatment unit. Br J Addict 87(1):125–127

Arora S, Roxburgh A, Bruno R, Nielsen S, Burns L (2013) A cross-sectional analysis of over the counter codeine use amongst an Australian sample of people who regularly inject drugs. Drug Alcohol Rev 32(6):574–581

Babalonis S, Lofwall MR, Nuzzo PA, Siegel AJ, Walsh SL (2013) Abuse liability and reinforcing efficacy of oral tramadol in humans. Drug Alcohol Depend 129(1–2):116–124

Baratta JL, Gandhi K, Viscusi ER (2014) Limited evidence that single-dose oral ibuprofen plus codeine is more effective for postoperative pain than either drug alone. Evid Based Nurs 17(2):51–52

Bateman C (2013) Is your prescribing serving a hidden addiction? S Afr Med J 103(6):359–361

Boyd CJ, McCabe SE (2008) Coming to terms with the nonmedical use of prescription medications. Subst Abuse Treat Prev Policy 3(1):22

Butler R, Sheridan J (2010a) Innocent parties or devious drug users: the views of primary healthcare practitioners with respect to those who misuse prescription drugs. Harm Reduction J 7(1):21

Butler R, Sheridan J (2010b) Innocent parties or devious drug users: the views of primary healthcare practitioners with respect to those who misuse prescription drugs. Harm Reduction J 7:21

Casati A, Sedefov R, Pfeiffer-Gerschel T (2012) Misuse of medicines in the European union: a systematic review of the literature. Eur Addict Res 18(5):228–245

[1]http://selfcare247.co.za/247-codeine-care-initiative/.

Chee E, Schneberger S (2003) British Colombia's pharmanet project. University of Western Ontario, London

Chetty R, Baoku Y, Mildner R, Banerjee A, Vallance D, Haddon A et al (2003) Severe hypokalaemia and weakness due to Nurofen(R) misuse. Ann Clin Biochem 40(4):422–423

Cooper RJ (2011) Over-the-counter medicine abuse—a review of the literature. J Subst Use 18(2):1–26

Cooper RJ (2011) 'Respectable addiction'—a qualitative study of over the counter medicine abuse in the UK

Cooper RJ (2013a) 'I can't be an addict. I am.' Over-the-counter medicine abuse: a qualitative study. BMJ Open 3(6):e002913

Cooper R (2013b) Surveillance and uncertainty: community pharmacy responses to over the counter medicine abuse. Health Soc Care Community 21(3):254–262

Dayer P, Desmeules J, Leemann T, Striberni R (1988) Bioactivation of the narcotic drug codeine in human liver is mediated by the polymorphic monooxygenase catalyzing debrisoquine 4-hydroxylation (cytochrome P-450 dbl/bufI). Biochem Biophys Res Commun 152(1):411–416

Dhital R, Whittlesea CM, Norman IJ, Milligan P (2010) Community pharmacy service users' views and perceptions of alcohol screening and brief intervention. Drug Alcohol Rev 29 (6):596–602

Dutch MJ (2008) Nurofen plus misuse: an emerging cause of perforated gastric ulcer. Med J Aust 188(1):56–57

Dyer BT, Martin JL, Mitchell JL, Sauven NC, Gazzard B (2004) Hypokalaemia in ibuprofen and codeine phosphate abuse. Int J Clin Pract 58(11):1061–1062

Ernest D, Chia M, Corallo CE (2010) Profound hypokalaemia due to nurofen plus and red bull misuse. Crit Care Resusc 12(2):109–110

Evans C, Chalmers-Watson TA, Gearry RB (2010) Combination NSAID-codeine preparations and gastrointestinal toxicity. NZ Med J 23(1324):92–93

Ferner RE, Beard K (2008) Over the counter medicines: proceed with caution. BMJ 336 (7646):7694–7696

Fleming GF, McElnay MJ, Hughes CM (2004) Development of a community pharmacy-based model to identify and treat OTC drug abuse/misuse: a pilot study. Pharm World Sci 26(5): 282–288

Frei M, Nielsen S, Tobin C, Dobbin M (2010) Serious morbidity associated with misuse of over-the-counter codeine-ibuprofen analgesics: a series of 27 cases. Med J Aust 193(5):294–296

French DP, James DH (2008) Reasons for the use of mild analgesics among English students. Pharm World Sci 30(1):79–85

Gourlay DL, Heit HA, Almahrezi A (2005) Universal precautions in pain medicine: a rational approach to the treatment of chronic pain. Pain Med 6(2):107–112

Hamer AM, Spark MJ, Wood PJ, Roberts E (2014) The upscheduling of combination analgesics containing codeine: the impact on the practice of pharmacists. Res Soc Adm Pharm 10 (4):669–678

Heather N, Bowie A, Ashton H, McAvoy B, Spencer I, Brodie J et al (2004) Randomised controlled trial of two brief interventions against long-term benzodiazepine use: outcome of intervention. Addict Res Theory 12(2):141–154

Horsfield E, Sheridan J, Anderson C (2011) What do community pharmacists think about undertaking screening and brief interventions with problem drinkers? Results of a qualitative study in New Zealand and England. Int J Pharm Pract 19(3):192–200

Hughes GF, McElnay JC, Hughes CM, McKenna P (1999) Abuse/misuse of non prescription drugs. Pharm World Sci 21(6):251–255

International Narcotics Control Board (2012) Narcotic drugs estimated world requirements for 2012. Comments on the reported statistics on narcotic drugs

Kahan M, Wilson L, Wenghofer EF, Srivastava A, Resnick A, Janecek E et al (2011) Pharmacists' experiences with dispensing opioids: provincial survey. Can Fam Phys 57(11):e448–e454

Khan N, Norman I, Dhital R, McCrone P, Milligan P, Whittlesea C (2013) Alcohol brief intervention in community pharmacies: a feasibility study of outcomes and customer experiences. Int J Clin Pharm 35(6):1178–1187

Kitabayashi Y, Ueda H, Narumoto J, Kita H, Nakamura K, Tsuchida H et al (2000) A case study of BRON (cough suppressant) tablet dependence–its social psychiatric and biological aspects. Japan J Alcohol Stud Drug Depend 35(5):295–305

Kolodny A (2013) Better late than never: time to up-schedule hydrocodone combination products. Pain Med 14(11):1627–1628

Kypri K, Langley JD, Saunders JB, Cashell-Smith ML, Herbison P (2008) Randomized controlled trial of web-based alcohol screening and brief intervention in primary care. Arch Intern Med 168(5):530–536

LA Gudin J (2013) The downside of upscheduling. Pain Med 14(11):1628–1629

Le Roux K (2013) Marketing of CPS. S Afr Pharm J 80(5):43–47

Leppert W (2011) CYP2D6 in the metabolism of opioids for mild to moderate pain. Pharmacology 87(5–6):274–285

Lessenger JE, Feinberg SD (2008) Abuse of prescription and over-the-counter medications. J Am Board Fam Med 21(1):45–54

MacFadyen LED, McGowan T (2001) Community pharmacists' experience of over-the-counter medicine misuse in Scotland. J R Soc Promot Health 121(3):185–192

Madadi P, Amstutz U, Rieder M, Ito S, Fung V, Hwang S et al (2013) Clinical practice guideline: CYP2D6 genotyping for safe and efficacious codeine therapy. J Popul Ther Clin 20(3):e369–e396

Manchia M, Alda M, Calkin CV (2013) Repeated erythromycin/codeine-induced psychotic mania. Clin Neuropharmacol 36(5):177–178

Matheson C, Bond C, Pitcairn J (2002) Misuse of over-the-counter medicines from community pharmacies: a population survey of Scottish pharmacists. Pharm J 269:66–68

Maxwell JC (2011) The prescription drug epidemic in the United States: a perfect storm. Drug Alcohol Rev 30(3):264–270

McAvoy BR, Tobin CL (2014) Fatal misuse of codeine-ibuprofen analgesics in Victoria, Australia. Med J Aust 200(3):150–151

McAvoy BR, Dobbin MD, Tobin CL (2011) Over-the-counter codeine analgesic misuse and harm: characteristics of cases in Australia and New Zealand. NZ Med J 124(1346):29–33

McBride AJ, Pates R, Ramadan R, McGowan C (2003) Delphi survey of experts' opinions on strategies used by community pharmacists to reduce over-the-counter drug misuse. Addiction 98(4):487–497

McCabe SE, Boyd CJ (2005) Sources of prescription drugs for illicit use. Addict Behav 30 (7):1342–1350

McDonough MA (2011) Misuse of codeine-containing combination analgesics. Med J Aust 194 (9):486

Medani S, Short S, Wall C (2010) Nurofen plus toxicity—what should nephrologists be alert to? Rare complications of an increasingly misused medication. NDT Plus. 2010 May 11

Medicines and Health Care Products Regulatory Agency (MHRA) (2009) Public assessment report. Codeine and dihydrocodeine-containing medicines: minimising the risk of addiction. MHRA, London (UK). Available from http://www.mhra.gov.uk/. Accessed 6 Oct 2013

Murnion B (2010) Combination analgesics in adults. Austr Prescriber 33:113–115

Myers B, Siegfried N, Parry CD (2003) Over-the-counter and prescription medicine misuse in Cape Town–findings from specialist treatment centres. S Afr Med J 93(5):367–370

Ng JL, Morgan DJ, Loh NK, Gan SK, Coleman PL, Ong GS et al (2011) Life-threatening hypokalaemia associated with ibuprofen-induced renal tubular acidosis. Med J Aust 194(6): 313–316

Nielsen S, Cameron J, Pahoki S (2010) Over the counter codeine dependence. Turning point alcohol and drug centre, Melbourne July 2010. Report No

Nielsen S, Cameron J, Lee N (2011) Characteristics of a non-treatment seeking sample of over-the-counter codeine users: implications for intervention and prevention. J Opioid Manag 7(5):363–370

Nielsen S, Tobin C, Dobbin M (2012) Over the counter codeine: examining the evidence for and against. Austr Pharmacist 31(3):236–240

Nielsen S, Cameron J, Pahoki S (2013) Opportunities and challenges: over the counter codeine supply from the codeine consumer perspective. Int J Pharm Pract 21(3):161–168

Nielsen S, Murnion B, Dunlop A, Degenhardt L, Demirkol A, Muhleisen P et al (2014) Comparing treatment-seeking codeine users and strong opioid users: findings from a novel case series. Drug Alcohol Rev 34(3):304–311

Nielsen S, Roxburgh A, Bruno R, Lintzeris N, Jefferson A, Degenhardt L (2015) Changes in non-opioid substitution treatment episodes for pharmaceutical opioids and heroin from 2002 to 2011. Drug Alcohol Depend 149:212–219

Nordmann S, Pradel V, Lapeyre-Mestre M, Frauger E, Pauly V, Thirion X et al (2013) Doctor shopping reveals geographical variations in opioid abuse. Pain Phys 16(1):89–100

Nuesch E, Rutjes AW, Husni E, Welch V, Juni P (2009) Oral or transdermal opioids for osteoarthritis of the knee or hip. Cochrane Database Syst Rev (4):CD003115

O'Donnell A, Anderson P, Newbury-Birch D, Schulte B, Schmidt C, Reimer J et al (2013) The impact of brief alcohol interventions in primary healthcare: a systematic review of reviews. Alcohol Alcohol 13:2013

Orriols L, Gaillard J, Lapeyre-Mestre M, Roussin A (2009) Evaluation of abuse and dependence on drugs used for self-medication: a pharmacoepidemiological pilot study based on community pharmacies in France. Drug Saf 32(10):859–873

Otto CCB, Lohrmann I, Zahradnik A, Bischof G, John U et al (2009) Brief intervention in general hospital for problematic prescription drug use: 12-month outcome. Drug Alcohol Depend 105 (3):221–226

Pates R, McBride A, Li S, Ramadan R (2002) Misuse of over-the-counter medicines: a survey of community pharmacies in a South Wales health authority. Pharm J 268:179–182

Phokeo V, Sproule B, Raman-Wilms L (2004) Community pharmacists' attitudes toward and professional interactions with users of psychiatric medication. Psychiatr Serv 55(12):1434–1436

Pilgrim JL, Dobbin M, Drummer OH (2014) Fatal misuse of codeine-ibuprofen analgesics in Victoria, Australia. Med J Aust 199(5):329–331

Reed K, Bond A, Witton J, Cornish R, Hickman M, Strang J (2011) The changing use of prescribed benzodiazepines and z-drugs and of over-the-counter codeine-containing products in England: a structured review of published English and international evidence and available data to inform consideration of the extent of dependence and harm. London: National Addiction Centre, King's College London, and School of Social and Community Medicine, University of Bristol

Robinson GM, Robinson S, McCarthy P, Camero C (2010) Misuse of over-the-counter codeine-containing analgesics: dependence and other adverse effects. NZ Med J 123 (1317):59–64

Roumie CLGM (2004) Over-the-counter analgesics in older adults, a call for improved labelling and consumer education. Drugs Aging 21(8):485–498

Roussin A, Bouyssi A, Pouche L, Pourcel L, Lapeyre-Mestre M (2013) Misuse and dependence on non-prescription codeine analgesics or sedative H1 antihistamines by adults: a cross-sectional investigation in France. PLoS ONE 8(10):e76499

Schatman ME, Darnall BD (2013) A practical and ethical solution to the opioid scheduling conundrum. J Pain Res 7:1–3

Shand F, Campbell G, Hall W, Lintzeris N, Cohen M, Degenhardt L (2013) Real-time monitoring of Schedule 8 medicines in Australia: evaluation is essential. Med J Aust 198(2):80–81

Sheridan J, Wheeler A, Ju-Hsing Chen L, Chen-Yun Huang A, Nga-Yee Leung I, Yow-Chyi Tien K (2008) Screening and brief interventions for alcohol: attitudes, knowledge and experience of community pharmacists in Auckland, New Zealand. Drug Alcohol Rev 27(4):380–387

Skurtveit S, Furu K, Borchgrevink P, Handal M, Fredheim O (2011) To what extent does a cohort of new users of weak opioids develop persistent or probable problematic opioid use? Pain 152(7):1555–1561

Sproule BA, Busto UE, Somer G, Romach MK, Sellers EM (1999) Characteristics of dependent and nondependent regular users of codeine. J Clin Psychopharmacol 19(4):367–372
Strobbe S (2014) Prevention and screening, brief intervention, and referral to treatment for substance use in primary care. Prim Care Clin Off Pract 41(2):185–213
Toms L, Derry S, Moore RA, McQuay HJ (2009) Single dose oral paracetamol (acetaminophen) with codeine for postoperative pain in adults. Cochrane Database Syst Rev (1):CD001547
UVUN (2011) The non medicinal use of prescription drugs. Discussion paper. Vienna
Vallejo RBR, Wang VC (2011) Pharmacology of opioids in the treatment of chronic pain syndromes. Pain Phys 14(4):E343–E360
Van Hout MC (2015) Nod and wave: an internet study of the codeine intoxication phenomenon. Int J Drug Policy 26(1):67–77
Van Hout MC, Bergin M, Foley M, Rich E, Rapca AI, Harris R et al (2014) A scoping review of codeine use, misuse and dependence, final report. Brussels: CODEMISUSED Project European Commission 7th Framework Programme
Vargas-Schaffer G (2010) Is the WHO analgesic ladder still valid? Twenty-four years of experience. Can Fam Physician 56(6):514–517
Wawruch MKM, Foltanova T, Ondriasova E, Luha J, Dukat A et al (2013) Characteristics of elderly patients who consider over-the-counter medications as safe. Int J Clin Pharm 35 (1):121–128
Wazaify M, Shields E, Hughes CM, McElnay JC (2005) Societal perspectives on over-the-counter (OTC) medicines. Fam Pract 22(2):170–176
Wazaify M, Hughes CM, McElnay JC (2006) The implementation of a harm minimisation model for the identification and treatment of over-the-counter drug misuse and abuse in community pharmacies in Northern Ireland. Patient Educ Couns 64(1–3):136–141
Wilk AI, Jensen NM, Havighurst TC (1997) Meta-analysis of randomized control trials addressing brief interventions in heavy alcohol drinkers. J Gen Intern Med 12(5):274–283
Wilsey B, Prasad H (2012) Real-time access to prescription drug monitoring databases. CMAJ 184(16):1767–1768
Wrobel JP (2003) Are we ready for the better medication management system? Med J Aust 178(9):448–450
Zahradnik A, Otto C, Crackau B, Löhrmann I, Bischof G, John U et al (2009) Randomized controlled trial of a brief intervention for problematic prescription drug use in non-treatment-seeking patients. Addiction 104(1):109–117
Zhou SF (2009) Polymorphism of human cytochrome P450 2D6 and its clinical significance: part I. Clin Pharmacokinet 48(11):689–723

Injection of Pharmaceuticals Designed for Oral Use: Harms Experienced and Effective Harm Reduction Through Filtration

Stuart McLean, Rahul Patel and Raimondo Bruno

Abstract Several pharmaceutical products are liable to 'abuse' or use outside their prescription, which frequently involves their injection. Examples are slow-release forms of morphine and oxycodone, and sublingual buprenorphine. During injection preparation, the drug is extracted into water, after crushing and heating the tablet if considered necessary. Since these products are designed for oral administration, they can contain excipients (ingredients other than the drug) which are poorly soluble, resulting in suspension of particles in the injection solution. Injected particles are able to produce medical complications such as the blockage of small blood vessels leading to ischaemia (inadequate blood flow) and tissue damage. Filtration can be used to remove particles from the suspension; including bacteria if the porosity is small enough (0.2 μm). However, filters are liable to blockage when overloaded, especially if the pore size is small. This problem can be minimised by using a larger pore size (e.g. 5–10 μm), but the resulting filtrate will contain many residual small particles. The use of two filters, coarse and fine, either sequentially or in a double membrane device, enables removal of the majority of particles as well as bacteria, although not quite meeting pharmaceutical standards for safe injection. Although not yet evaluated by a clinical trial, this highly effective filtration process would be expected to greatly reduce the risk of vascular and related complications, as well as non-viral infections. Careful technique ensures that drug is not lost by filtration, a priority for most drug consumers. Practical issues that affect acceptability of filtration by injecting drug users, including ease of use and cost, will need to be considered. However, given the laboratory evidence demonstrating the effectiveness of filters it is time to consider these tools as essential for safe injection as sterile needles/syringes for the world's approximately 16 million people who inject drugs.

Keywords Non-Medical Use of Prescription Drugs · Injecting drug use · Filtration · Harm reduction

---

S. McLean · R. Patel · R. Bruno (✉)
School of Medicine, University of Tasmania, Hobart, Australia
e-mail: raimondo.bruno@utas.edu.au

Curr Topics Behav Neurosci (2017) 34:77–98
DOI 10.1007/7854_2016_470
Published Online: 02 April 2017

Contents

1 Introduction: Illicit Injection of Pharmaceuticals

Pharmaceutical drug products have some advantages over illicit street drugs when used for non-medical purposes, in that the drug substance and dose are known, and potentially toxic contaminants are absent. However, when pharmaceuticals that are designed for oral administration (tablets, capsules, sublingual films, etc.) are used to prepare solutions for injection, there is a great potential for harm. In addition to adverse effects from the drug itself, intensified by the high blood levels that follow intravenous injection, other constituents in oral dosage forms can cause medical complications. These constituents, termed excipients, are considered to be pharmacologically inert and are used to facilitate manufacture of the dosage form, maintain its stability, and aid or control the release of active drug in the gastrointestinal tract (Rowe et al. 2012). Examples of excipients are diluents (lactose, talc—a hydrated magnesium silicate); binders and disintegrants (starches, cellulose derivatives); glidants and lubricants (silica compounds, magnesium stearate); and coatings (polymers). Colourings, flavourings, preservatives, antioxidants and emulsifiers are also used. Recently, sustained-release oral formulations are made in which the drug is slowly released from a complex polymer and wax matrix (Miller et al. 2008). These are generally considered safe when used as intended [although some people experience sensitivity reactions (Haywood and Glass 2011)], but when the pharmaceutical dosage forms are used to make injections, the excipients can produce large numbers of poorly soluble particles which give rise to cardiovascular and other medical complications. A number of excipients of the more commonly injected pharmaceutical products, and their properties relevant to injection-related harms, are summarised in Table 1.

The medical complications of injecting illicit pharmaceuticals are many and varied, and include skin and soft tissue injuries; pulmonary, cardiac and vascular conditions; and local and systemic infections (Chiang and Goldfrank 1990; Cherubin and Sapira 1993; Dwyer et al. 2009). Most of these can be traced to the nature of the non-drug substances in the injection, and the manner and frequency of injections used. Hypodermic needles breach the normal protective barrier of the skin and can introduce to the underlying tissues substances which are otherwise excluded. The biological effects of abnormal particles depend on the site of their injection and the tissue involved (e.g. skin, muscle, vein, artery, etc.); the number and size distribution of the particles and their chemical composition (e.g. silica, polymer, etc.); and the frequency and duration of injecting behaviour. The mechanisms of particle-induced reactions result from irritant damage at the sites of injection and their deposition in blood vessels and other tissues, associated inflammatory responses, and ischaemia due to blockage of blood vessels. These adverse reactions are commonly exacerbated by concomitant bacterial or fungal infections associated with non-sterile injections.

In order to understand the range of adverse events that can arise from the introduction of insoluble particles in the bloodstream it is important to consider the events that occur when a drug is injected. This chapter aims to provide a review of injection-related harms and the evidence for filtration as a means of reducing these harms.

2 Thrombosis and Inflammation

The body has powerful defence systems which can be activated by the injection of crushed tablets (Kumar et al. 2005). These are blood coagulation (thrombosis) and inflammation, which have evolved to prevent blood loss and destroy microorganisms and other foreign material, but they can also cause collateral damage to the host. Thrombosis and inflammation can occur together, leading to a complex pathological process, especially with repeated exposures such as with prolonged injecting behaviour (Del Giudice 2004).

Any disturbance of the normal circulation of blood can initiate the clotting process, leading to the formation of an intravascular clot, or thrombus (Kumar et al. 2005). Disturbances can be produced by the presence of foreign material, damage to the vessel wall, or when blood flow becomes slow or turbulent. Formation of a thrombus narrows the vessel and reduces the blood flow, which may cease if the thrombus is large enough.

Foreign material including microorganisms can stimulate an acute inflammatory response, which involves the migration of white blood cells (leukocytes) to the site and their attempt to engulf (phagocytose) the offending agents (Kumar et al. 2005). Leukocytes release enzymes able to destroy microorganisms but which are also

Table 1 Common excipients in pharmaceuticals injected by people who inject drugs, and their relevant properties

	MS Contin®	Kapanol®	Oxycontin OC®	Subutex®	Suboxone® tablet	Suboxone® film	Particle size	Solubility in water	Response to heat
Hydroxy/ ethylcellulose	✓	✓					Varies, <177 μm	Soluble	Softens >130 °C
Hypromellose	✓	✓	✓			✓	Varies	Soluble in cold water, forms viscous colloidal solution; insoluble in hot water	Browns at 190 °C
Lactose (anhydrous)	✓		✓	✓	✓		Varies, 80–93% >45 μm	Soluble	MP >220 °C
Magnesium stearate	✓		✓	✓	✓		Fine greasy powder	Practically insoluble	Varies, MP >117 °C
Maize starch		✓		✓	✓		2–32 μm, average 13 μm	Practically insoluble in cold, soluble >71 °C	Swell 64 °C, gelatinize 71 °C
Mannitol				✓	✓		Varies, avg particle diameter 250–520 μm	1 in 5.5	MP 166 °C
Povidone			✓	✓	✓		Varies, 95% >50 μm	Viscous solution	Softens 150 °C
Stearyl/cetostearyl alcohol	✓		✓				'flakes'	Practically insoluble	Varies, MP >48 °C
Talc	✓	✓	✓				>99% less than 74 or 44 μm	Practically insoluble	<7% loss on ignition

Source Rowe et al. 2012

capable of causing local tissue damage. This defensive system works better for infectious agents than for tablet particles, such as talc, which are generally too large and inert to be eliminated in this way. In this case the inflammation progresses to a chronic stage in which the major inflammatory cells are large leukocytes called macrophages, and repair is attempted by fibroblast cells which replace damaged tissue with connective (scar) tissue. Eventually the foreign body can become enclosed by macrophages (which become transformed into epithelium-like cells and giant cells), fibroblasts and connective tissue, to become a lump termed a foreign body granuloma. Inside blood vessels, granulomas can develop in association with a thrombus.

3 Embolism and Ischaemia

The heart pumps blood through two circulations, the right side through the lungs (pulmonary circulation) and the left side to the rest of the body (systemic circulation) (Seeley et al. 2006). In each circulation the blood passes through vessels of progressively decreasing diameter: arteries (large and small), arterioles and capillaries, and then returns through ever-widening vessels: venules, small to medium and large veins, and back to the heart. Medium-sized arteries are 40–300 μm in diameter, arterioles are 9–40 μm, and capillaries 7–9 μm. In the capillaries there are exchanges between blood and tissues. The pulmonary capillaries excrete carbon dioxide and collect oxygen, and the systemic capillaries deliver oxygen and nutrients and remove wastes.

Injected particles tend to lodge in vessels which are smaller than themselves, causing an obstruction or embolism (Kumar et al. 2005). The site of this depends on the size of particles and where they were introduced into the circulation. Injections are usually most readily made into the prominent superficial veins on the forearms, after which particles will only encounter narrower vessels after they pass through the right side of the heart and enter the pulmonary circulation. Studies with dogs injected intravenously with radiolabelled polymer microspheres found that size determined their tissue distribution (Kanke et al. 1980). Within a few minutes particles of all sizes (3–12 μm) were rapidly taken up by the lungs. The 12 μm particles remained indefinitely within the pulmonary vasculature, while smaller particles were subsequently washed out of the lungs to be eventually removed from circulation by macrophages in the liver and spleen. This redistribution was complete and rapid (<24 h) for 3 and 5 μm particles, but incomplete after 4 weeks for the 7 μm particles. There was evidence of agglomerates of from 3 to 10 microspheres within the vasculature, possibly via platelet adhesion. In a similar study in rabbits, it was found that microspheres (10^6/dose) of 15.8 μm were trapped by the lungs within 1–2 min, while those of 1.3 μm accumulated in the liver and spleen (Illum et al. 1982). There was no evidence of adverse effects even after injection of larger particles (40–160 μm), but 30 μm fibres of the same material were rapidly fatal, with signs of acute pulmonary embolism. This and other studies (Decuzzi et al. 2009) show that the biological effects of insoluble particles depend on their shape as well as size.

4 Injection Site and Blood Vessels

Most people who inject drugs (PWID) non-medically prefer intravenous injection because of the rapid and intense pharmacological effect. This is a consequence of the rapid administration of a bolus of drug solution which produces a high initial concentration in the blood and subsequently in the brain (Shargel et al. 2005). Initial injections are typically made into the readily accessible superficial vein in the crook of the arm or cubital fossa (Darke et al. 2001). The walls of all blood vessels are complex, multilayered structures (Seeley et al. 2006). In veins, the innermost layer (the tunica intima) comprises a sheet of flat cells (the endothelium), and a thin layer of connective tissue and elastic fibres. This in turn is surrounded by successive layers of smooth muscle and connective tissue, and the walls are supplied with their own blood vessels and nervous system.

Pharmaceutical injections are formulated to minimise insult to delicate vascular tissue. These injections are pH-buffered, isotonic with body fluids, filtered and sterile, and are usually administered quite slowly (British Pharmacopoeia Commission 2005; Boylan et al. 1996). Rapid injection of a crushed tablet suspension effectively produces a physical and chemical assault on the vein, eliciting an inflammatory response in the vessel wall (phlebitis), commonly associated with thrombus (thrombophlebitis) (Del Giudice 2004). Damaged tissue can be repaired, but when there is continued exposure to irritants normal tissue is destroyed and replaced by fibrous connective tissue, with a consequent loss of normal vessel structure and function (Kumar et al. 2005). For example, impaired flow in superficial veins causes accumulation of fluid (oedema) and predisposes to skin infections and ulcers.

With intravenous injections there is a risk of extravasation, where some or all of the injection is made outside the vein, and this becomes more likely after the vein has been damaged by previous injections (Del Giudice 2004). Irritant injections deposited in subcutaneous tissue are not readily cleared by the body and cause chronic inflammation, formation of a granuloma, and sometimes tissue death (necrosis). Destruction of skin and underlying tissues results in the formation of an ulcer or, with infection, an abscess. At all stages in this complex pattern of events, the presence of microorganisms leads to infection which can greatly worsen the clinical situation. Abscesses and chronic wounds are common amongst injecting drug users (Del Giudice 2004; Smith et al. 2015).

As local damage makes it harder to inject into a vein, other sites are used, in the arm, hand, lower limbs, neck, groin and digits (Darke et al. 2001). After the loss of accessible veins, extravascular injections may be made into subcutaneous or muscle tissue (Del Giudice 2004). Injections into arteries in the limbs, which can occur accidentally when probing for a vein, or deliberately after other sites become obliterated, are extremely harmful when particles are involved. Particles large enough to lodge in the arterial system (emboli) reduce the blood flow causing ischaemic damage and necrosis in downstream tissue. The consequences can be gangrene and amputation of affected digits or the limb itself. Injections of crushed buprenorphine tablets into the radial, brachial, or femoral artery can produce an immediate burning pain, which continues for days, with swelling and necrosis in the distal tissue (variously, hand, foot, or pubic area) (Del Giudice et al. 2005). The same tablets injected into the carotid artery have been associated with neurological symptoms and brain lesions seen on magnetic resonance imaging, consistent with acute cerebral infarction from a shower of microemboli (Lim et al. 2009).

5 Pulmonary System

After intravenous injection the entire dose moves in the systemic venous system to the right side of the heart and then into the pulmonary circulation. Thus, the right heart and its valves and the pulmonary vessels and lungs are major sites for injection-related harms (Cherubin and Sapira 1993; Gotway et al. 2002). The lungs effectively act as a filter which removes most of the insoluble particles which pass through the heart. Only particles smaller than about 7 μm can normally pass through pulmonary capillaries, and these then return to the left side of the heart and enter the general systemic circulation where they can reach all other tissues. In addition, the pulmonary blood flow can be shunted around obstructed capillaries, allowing larger particles to reach the systemic circulation. Talc particles have been found in kidneys and lymph nodes, and seen in the retinal vessels (Pare et al. 1979; Schoenberger and Agarwal 2013). Most small particles will eventually be sequestered by the macrophages in the liver and spleen (Kanke et al. 1980; Illum et al. 1982).

Much of what we know about the interaction in humans between injected particles and the pulmonary system comes from histological examination of tissue samples taken as surgical biopsies or at autopsy. Examination by light and electron microscopy can distinguish the types of particle [e.g. microcrystalline cellulose (Ott et al. 2003), talc (Pare et al. 1979), crospovidone (Ganesan et al. 2003), cornstarch (Schneider et al. 2010)] and identify the pathological responses (Sigdel et al. 2011). Clinical notes together with other evidence of injection drug use (injection site skin lesions, drug analyses) can be used to investigate associations between these pathological findings and illicit drug use.

Autopsies on 70 intravenous drug users showed foreign material admixed with thrombus in the pulmonary arteries and arterioles (Tomashefski and Hirsch 1980). The foreign material included talc (5.8–23 μm), microcrystalline cellulose (10–250 μm), cornstarch and cotton fibres, the latter probably from a cotton filter. These deposits were only associated with injection of crushed tablets and not illicit heroin as this evidently contained less insoluble matter. Following the injection of pentazocine tablets, one lung had an enormous amount of cellulose, which was even visible macroscopically on the cut lung surface.

The pulmonary arterial blood flow can be occluded by embolic particles and the granulomas which form around them (Tomashefski and Hirsch 1980; Arnett et al. 1976). The obstruction to flow can raise the pulmonary arterial pressure (pulmonary hypertension) which in turn increases the load on the right ventricle. Over time, both the pulmonary arteries and right ventricle develop thicker and more muscular walls (hypertrophy) in response. However, the increased work required of the right ventricle can lead to its failure. Even in patients not previously suspected of injecting tablets, intravascular talcosis has been found to be associated with dyspnoea, pulmonary fibrosis and hypertension, and congestive heart failure (Griffith et al. 2012).

Particles can slowly migrate through the vascular wall into interstitial lung tissue, forming granulomas and non-functioning scar tissue or fibrosis, which can be indicative of long-term drug use (Tomashefski and Hirsch 1980; Arnett et al. 1976). Conversely, cornstarch particles can be removed by metabolism and their presence indicates more recent injection (Low and Nicol 2006). Although small particles can escape pulmonary capillary filtration, particles of talc (≥ 5 μm) and titanium dioxide (<0.5 μm) small enough to pass through capillaries have been found deposited in lung tissue (Dettmeyer et al. 2010).

Unlike a major pulmonary embolism event, multiple small vessel emboli are clinically silent, and the damage which occurs after each injection may produce no symptoms (Arnett et al. 1976). In rodents, the lethal intravenous dose of embolic microspheres depended on the number injected and was lower for those with larger diameter (range 13–90 μm), which produce a greater ischaemic region (Davis and Taube 1978). The effects of repeated injections of small particles accumulate until the damage has become extensive and symptoms slowly develop, progressing from exertional dyspnoea, cough and chest pain to pulmonary hypertension and congestive heart failure (Arnett et al. 1976; Jorens et al. 2009). A recent review of 373 cases of foreign body pulmonary embolism from autopsies during 1997–2013 in NSW (Darke et al. 2015) found that in most cases (2/3) death was due to drug overdose, indicating that pulmonary disease is responsible for more morbidity than mortality. A variety of other thoracic complications of injecting drug use have been described (Gotway et al. 2002).

6 Infections

Infections have long been recognised to be among the most serious complications of injecting drug use (Cherubin and Sapira 1993; Gordon and Lowy 2005; Kaushik et al. 2011). The most common bacteria causing non-pulmonary infections are *Staphylococcus aureus* and *Streptococci*, often commensals from the injector's own skin (Bassetti and Battegay 2004). However, many other bacteria as well as fungi (*Candida* and *Aspergillus*) have been found in PWID (Cherubin and Sapira 1993; Gordon and Lowy 2005; Kaushik et al. 2011; Orangio et al. 1984; Ebright and Pieper 2002). The handling and equipment required to prepare an injection from an oral dosage form increases the risk of contamination and inoculation with bacteria. Oral bacteria (*Streptococci* and anaerobes) can come from crushing the tablet in the teeth and licking the needle to clean it or test the strength of the drug.

Microorganisms from the skin, mouth and utensils can be injected in non-sterile drug preparations. In addition, the presence of tissue damage, ischaemia, and necrosis all provide a favourable environment for infection to become established (Gordon and Lowy 2005; Kaushik et al. 2011; Ebright and Pieper 2002). The most vulnerable areas include the injection site and surrounding tissue, and the heart and lungs as organs of first contact.

Skin and soft tissue infections (skin abscesses and cellulitis) are the most common reasons for hospital admissions in people who inject drugs (PWID) (Bassetti and Battegay 2004; Ebright and Pieper 2002). These infections are usually abscesses around the site(s) of injection. Injections made into the skin (subcutaneous) or muscle (intramuscular), known as "skin popping", carries a high risk of abscesses, although not with normal aseptic medical practice (Murphy et al. 2001). Complications, especially when the circulation is compromised, include chronic skin ulcers on the legs and feet, which can be large and debilitating and heal only slowly (Ebright and Pieper 2002).

Vascular infections include endocarditis (70% involving the right-side tricuspid valve), septic thrombophlebitis and sepsis (Gordon and Lowy 2005). Injected particles can damage the normally resistant layer of endothelial cells that line the vasculature, heart and its valves, making it more susceptible to infection by microbes introduced into the circulation (Sousa et al. 2012; Saydain et al. 2010). The microbes (usually *Staphylococci* or *Streptococci* but sometimes *Candida*) form bulky, friable growths (vegetations) which include thrombus and inflammatory cells. Complications include a generalised blood infection (sepsis), lung and brain embolization, and respiratory and cardiac failure. Although the overall incidence of infectious endocarditis has been declining in the general population, it is increasing amongst injecting drug users, evidently because of the associated risk factors (Tung 2015).

Injecting drug users have a tenfold increased risk of community-acquired pneumonia, most frequently from *Staphylococci* species and *Haemophilus influenzae* but occasionally from *Candida albicans* attributable to the use of lemon juice to acidify injections (Cherubin and Sapira 1993; Gotway et al. 2002, Hind 1990). Septic emboli from the injection site or vegetations on the tricuspid valves can lodge in the lungs and produce infection. Disseminated infections, commonly streptococcal, cause septicaemia with high fever, toxicity, and renal failure (Cherubin and Sapira 1993). Musculoskeletal infections, including septic arthritis and osteomyelitis, are often polymicrobial or anaerobic, involve the vertebral spine or uncommon places such as the sacroiliac joint, and may be indolent with only pain as a symptom (Gordon and Lowy 2005).

7 Filtration as a Harm Reduction Measure

In hospital practice, where solutions of drugs, salts and nutrients are extemporaneously prepared for intravenous administration to individual patients, filtration is routinely used to remove particles and cellular microorganisms (Kuramoto et al.

2006). Solutions passed through a 0.2 μm filter are considered to be sterile, as this porosity removes bacteria and fungi although not viruses or prions, which however are only likely to be present through contamination by infected blood, a risk which is avoidable in hospital practice although common amongst people who inject drugs illicitly (Caflisch et al. 1999).

Filtration therefore offers the prospect of removing particles and microoganisms from injections prepared from oral and topical pharmaceuticals, and thereby greatly reducing the injection-related medical complications. Several independent studies have demonstrated the potential for reduction of particulate burden from injection of tablet pharmaceuticals (Roux et al. 2011; Patel et al. 2012; McLean et al. 2009; Scott et al. 1998) and other illicit drugs (Scott 2005), as well as the reduction of bacterial contamination (Caflisch et al. 1999; Ng et al. 2015). However, there are several problems that have prevented this from becoming routine. One is that injections prepared from crushed tablets contain a very heavy burden of particles, which quickly block all but the coarsest filters (McLean et al. 2009; Scott et al. 1998), as the majority of commercially available syringe filters are designed for hospital practice with largely uncontaminated solutions. This reinforces the argument for effective filtration to remove the harmful particles. In response, many consumers revert to use of coarse filters [such as cigarette filters (Patel et al. 2012; Scott et al. 1998)], which are less effective at removing particulate contamination but avoid loss of dose from filter blockage. Some commercially available filters have been designed specifically for use by people who inject drugs (such as the 10 μm Sterifilt®), and provide superior particle removal while minimising blockage (Roux et al. 2011; Keijzer and Imbert 2011). An alternative approach is to remove large particles through use of a coarse filter, and then follow this with a bacterial filter (Ng et al. 2015). This increases the time and number of steps required for injection preparation, which may be a barrier to use (Keijzer and Imbert 2011); and single device dual-membrane syringe filters (e.g. 5.0/0.22 or 0.8/0.2 μm filters) may address this challenge. A second barrier to take-up is that the filtration procedure, in order to be acceptable to consumers, must not result in a loss of drug (Caflisch et al. 1999), and this requires scientific verification. Lastly, each drug formulation may differ significantly from others in terms of their insoluble particulate burden and the characteristics of the tablet excipients (see Table 1). As such, filtration methods require validation on a range of formulations which are commonly used to prepare injections. In addition, not all of the filter types are routinely available internationally at places where PWID access sterile injecting equipment, and cost may be an additional barrier to take-up. Given these issues, our group has examined the effectiveness of filtration on removal of both particulate and bacterial contamination from a range of pharmaceutical products used by PWID, using both makeshift and commercially available filters, as well as examining the recovery of active drug in the filtration process. These studies [both previously published (Patel et al. 2012; McLean et al. 2009; Ng et al. 2015) and unpublished work] are summarised in the next section.

8 General Filtering Study Methodology

In order to produce practically generalizable results from these studies, our group have adopted the following approach: initially, identify the types of pharmaceuticals used among PWID and how they are filtered; interview consumers in relation to the procedures that they use to prepare these drugs for injection; replicate these methods in the laboratory and determine the resultant particulate and bacterial reduction, and recovery of active drug, using a range of filter devices.

8.1 Survey of People Who Inject Drugs

Interviews with PWID were conducted as part of the Illicit Drug Reporting System (IDRS), a surveillance system for illicit drug markets across Australia. Participants (n = 898) were recruited through a purposive sampling strategy in each capital city of Australia in 2014, with trained interviewers conducting face to face interviews. Inclusion criteria required age greater than 18 years and at least monthly injection of drugs in the past 6 months. Participant mean age was 41 (range 18–67), the majority were male (69%), and unemployed (83%), with a mean of 10 years of school education, and 46% had trade qualifications. Almost all (98%) were injecting weekly or more frequently, with 45% were injecting daily. Full methodological detail is available elsewhere (Stafford et al. 2014).

8.2 Drug Extraction and Filtration

Qualitative interviews were conducted with PWID in order to determine the standard practices used by consumers to prepare tablets for injection. These are detailed elsewhere (Patel et al. 2012; McLean et al. 2009). In general, preparations may be considered as a cold-extraction method or a hot-extraction method. In both methods, tablets are wiped with alcohol swabs to remove wax coating and any colouring, the tablet is crushed, and sterile water is added to the resultant powder. In the cold-extraction method, the mixture is stood for a short time with occasional stirring to allow the drug to dissolve, although this will not be visible in the presence of much insoluble material. In the hot-extraction method, heat is applied to speed up the dissolution, and this can liquefy any wax constituents which are then liable to carry through to the injection.

Solutions were either analysed unfiltered, analysed following filtration using cigarette filters (Stuart Alexander & Co., New South Wales, Australia), commercially available 0.45 and 0.22 μm syringe filters (Millipore, New South Wales, Australia) or combination 0.80/0.20 μm syringe filters (PALL Life Sciences, Michigan, USA).

The process for filtration using a cigarette filter was that the filter was halved, placed into the drug mixture, and allowed to soak up the solution. Using a 5 mL syringe, the solution was drawn up completely. Subsequently, sterile water was added to the mixing vessel as a rinse, and also drawn up into the syringe. Using 0.45 and 0.22 μm syringe filters, the solution *following* the initial cigarette filtration was passed through the syringe filter and rinsed with water. Using the combination 0.80/0.20 μm filter, the initially prepared solution was drawn through the filter without the initial step of cigarette filtration.

8.3 Analysis of Particle Content

This process is detailed in full elsewhere (McLean et al. 2009). Briefly, a 20 μl aliquot of each sample was pipetted to a clean glass microscope slide. Slides were viewed under a light microscope with eyepieces that showed a rectangular area and a linear scale, calibrated using the ruled lines of a Neubauer blood cell counting chamber. Photographs of the slides were taken and particles were counted in size groups, based on counts from five fields within each slide using standardised counting rules. Total particles were estimated from average count per field multiplied by the ratio of field area to coverslip area, and then scaled up to the volume of injection to give the number of particles in the injection mixture. Counts were based on means of three replicate preparations; and are reported here as proportionate particulate load in comparison to the load for unfiltered solutions.

8.4 Analysis of Active Drug Content

Solutions resulting from each filtration technique were analysed for target drug content using either high performance liquid chromatography (morphine, buprenorphine) or UV spectrophotometry (oxycodone). Full examples of each methodology are provided elsewhere (Patel et al. 2012; McLean et al. 2009). Amounts of active drug recovered were based on triplicate analyses, calibrated against standards, and reported as a proportion of the stated content of each tablet.

8.5 Drugs Examined

Tablets examined were based on those used among PWID. These included 60 mg morphine MS Contin® tablet (Mundipharma); 100 mg morphine Kapanol® tablet (Mayne Pharma); 80 mg oxycodone OxyContin® OC tablet (Mundipharma); 8 mg buprenorphine Subutex® tablet (Reckitt Benckiser); 8 mg (buprenorphine)/2 mg (naloxone) Suboxone® tablet (Reckitt Benckiser); and 8 mg (buprenorphine)/2 mg (naloxone) Suboxone® sublingual film (Reckitt Benckiser).

8.6 Bacterial Suspension

The commercial syringe filters were designed for medical applications with low contamination, and in light of the high particulate count in these solutions, it was not known if these devices would maintain their ability to remove relevant bacteria. As such, solutions were prepared using the most challenging tablet for filtration (MS Contin) and three bacterial strains noted as common contributors to infection in PWID (Gordon and Lowy 2005; Kaushik et al. 2011): *S. aureus* (ATCC 29213), *Streptococcus pyogenes* (SP; ATCC 19615) and *Pseudomonas aeruginosa* (PAO, reference strain PAO1). Full details of growth methodology, concentration determination and bacterial enumeration are available elsewhere (Ng et al. 2015). Briefly, concentrations of bacteria were retrospectively determined by serial dilution, with concentrations of each bacterial species ranging from 1.5 to 8.6×10^6 colony-forming units (CFU) mL^{-1}. Bacterial suspensions were plated on brain heart infusion broth solidified with 1.5% bacteriological agar (BHIA, Oxoid CM1135 and LP0011). Aliquots of each dilution were incubated for 24 h. The number of colonies of each bacterial species was determined by macroscopic examination, and replicated in triplicate.

9 Results

9.1 Filtration Methods Used by PWID

Among the Australian national sample of PWID, injection of pharmaceuticals was a common behaviour, in particular injection of morphine (predominantly MS Contin® tablets) and of oxycodone (predominantly OxyContin OC®). Respondents were also asked for their preparation process on the last time they injected each pharmaceutical (Table 2). Noteworthy was that the majority reported some degree of filtering (more than 95% in the case of the most common tablets; more than 90% for buprenorphine tablets; but around 80% for buprenorphine film). This clearly suggests that filtration as a general process is acceptable to consumers. However, it is also noteworthy that the vast majority of filtration was conducted using makeshift filters (cotton balls, cigarette filters) rather than the more effective syringe filters (typically less than 15%). This may be reflective of access issues: commercial syringe 'wheel' filters are not standard equipment across all Australian Needle and Syringe Program outlets, although many services do aim to stock them; and in many cases outlets are required to charge for their provision (typically $2–3 AUD)

Table 2 Injection of pharmaceuticals and processes for preparation among people who frequently inject in Australia, 2014 ($N = 898$)

	Morphine tablet	Oxycodone tablet	Subutex®	Suboxone® tablet	Suboxone® film
Injected in past 6 months	34.7%	27.2%	12.0%	3.8%	8.2%
Preparation method on last occasion of use					
Cold extraction	22.3% ($n = 69$)	25.1% ($n = 60$)	85.0% ($n = 91$)	79.4% ($n = 27$)	84.7% ($n = 61$)
Hot extraction	77.7% ($n = 240$)	74.9% ($n = 179$)	15.0% ($n = 16$)	20.6% ($n = 7$)	15.3% ($n = 11$)
Filter used on last occasion of use					
No filter	3.5% ($n = 11$)	1.2% ($n = 3$)	9.2% ($n = 10$)	8.8% ($n = 3$)	18.1% ($n = 13$)
Cotton	18.1% ($n = 56$)	32.4% ($n = 78$)	27.5% ($n = 30$)	26.5% ($n = 9$)	25.0% ($n = 18$)
Cigarette	60.3% ($n = 187$)	53.1% ($n = 128$)	40.4% ($n = 44$)	38.2% ($n = 13$)	47.2% ($n = 34$)
Wheel	13.9% ($n = 43$)	11.6% ($n = 28$)	12.8% ($n = 14$)	23.5% ($n = 8$)	6.9% ($n = 5$)

Source (Stafford et al. 2014)

which may create a barrier to use in comparison to the substantially cheaper cigarette filters. A second issue may relate to practical issues in use of the commercial wheel filters, which will be discussed below. Also notable in these results is that while the buprenorphine products were typically extracted without heating, the more commonly used morphine and oxycodone tablets were predominantly extracted using heat. Given that the process of heating interacts with the characteristics of the excipients in each tablet, this has important implications for the effectiveness of the filtration process, as discussed below.

9.2 *Particulate Contamination Following Filtration*

1. *Blockade of filters*

The commercial syringe filters most readily available to PWID in Australia are 0.45 and 0.22 μm. However, these are not commonly utilised for morphine and oxycodone by consumers. In our laboratory, assessments of these filters were made for both MS Contin® (McLean et al. 2009) and OxyContin® (Patel et al. 2012), and also for all buprenorphine formulations (both tablets and sublingual film). After cold extraction, the filters were blocked by the large degree of particulate contamination, which provided substantial resistance and rendered complete filtration impossible even with substantial pressure applied. Using a hot-extraction process allowed some solutions to pass successfully through 0.45 μm but not 0.22 μm filters (McLean et al. 2009), however this process comes at a cost to effectiveness of the filtration process (discussed below). As such, in our laboratory studies we examined a two-step filtration process, whereby extracts were initially drawn up through a cigarette filter and then passed through a 0.45 or 0.22 μm syringe filter. This allowed filtration to be completed in all cases. The double membrane filter (0.8/0.2 μm) allowed complete filtration of both cold and hot extracts. It is notable that this is not the case for all double membrane filters—in our tests of 5.0/0.2 μm filters (data not shown) these have become easily blocked when filtering buprenorphine, requiring substantial time (>180 s) or excess pressure to filter. Where excessive pressure was applied, while the filtration process was quicker (<60 s) there was a substantial increase in particulate contamination in the filtrate, suggesting that the membrane had been ruptured in this process.

2. *Influence of hot extraction*

As noted in Table 2, the majority of PWID appear to extract their morphine and oxycodone tablets using hot extraction. Not only is this a quicker extraction process, but we have demonstrated that this substantially reduces the particulate burden for both tablets (Patel et al. 2012; McLean et al. 2009). The heating and subsequent cooling resulted in an agglomerated waxy mass, typically too large to be taken up by the syringe. While this makes the mechanics of the filtration process faster and simpler, it also introduces new challenges. As noted in Table 1, many of the tablets that are injected contain excipients with relatively low melting points, and in addition the heating process can change the solubility of some excipients. If the solution is still warm when passed through the filter, some excipients may pass through the filter and resolidify in the syringe or the body. The effect of this is plainly apparent in Fig. 1 where the 0.45 and 0.22 μm filters are mostly less effective in removing large (≥ 10 μm) particles for hot-extracted solutions in comparison to cold-extracted solutions, despite these particles being substantially larger than the rated size of the filter and that the filtering process itself was smooth and did not require substantial pressure.

3. *Effectiveness of filters on removing particulate contamination*

Across all products studied, cigarette filters typically reduced the particulate burden of large particles (≥ 10 μm) by approximately 50%, but typically provide limited benefit for removal of particles smaller than 10 μm (Fig. 1). While pore sizes vary, typically cigarette filters range between 20 and 40 μm (Sameer 2010), which allow substantial scope for particles large enough to cause harms to successfully pass through the filtering process. Due to the handling process involved in cutting and placing cigarette filters, the raw numbers of particles <10 μm were often greater following cigarette filtration than an unfiltered solution (data not shown). Following a cigarette filtration step with an 0.45 or 0.22 μm filter removed on average 95% or more of both ≥ 10 and <10 μm particles. However, as noted above, this process was much less successful for hot-extracted solutions, with 75–85% of ≥ 10 μm particles removed in this process. Lastly, the single device double

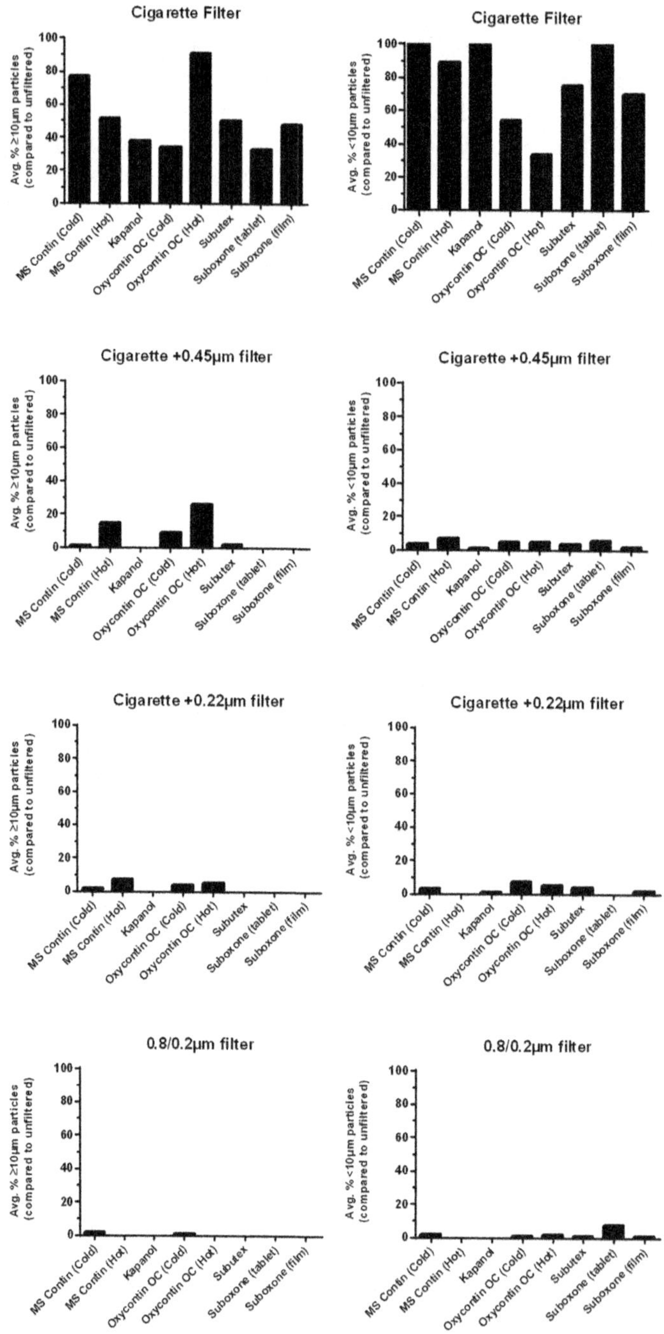

Fig. 1 Summarised results of filtration studies

membrane (0.8/0.2 μm) filter was easier and quicker to work with and successfully removed 99% of ≥ 10 μm particles and typically 98% of <10 μm particles, with the notable exception of suboxone tablets. These tablets have a particularly high load of <10 μm particles, and it is possible that this excess burden may have contributed to a rupture of the filter membrane in the process. Lastly, while these methods demonstrate the effectiveness of filters in dramatically reducing particulate contamination, none of the methods met pharmaceutical standards for injection [no more than 3000 particles >10 μm and no more than 300 particles >25 μm (British Pharmacopoeia Commission 2005)].

4. *Effectiveness of filters on removing bacterial contamination*

Given that the results of the particulate counting studies demonstrate that it is possible for particles larger than the rated filter pore size to pass, it was important to determine the success of the filters at removing small bacteria when examined in conjunction with a prepared tablet. In this study, cold-extracted MS Contin® tablets were chosen given that these produced the solutions that were most difficult to work through the filtration process. As shown in Fig. 2, cigarette filters were ineffective at bacterial removal, while syringe filters, in particular the 0.22 μm filters, effectively removed *S. aureus*, *S. pyogenes* and *P. aeruginosa* to levels beneath the limit of quantification, even in the presence of highly contaminated solutions (Ng et al. 2015).

5. *Recovery of active drug*

Concerns in relation to loss of active drug are an identified barrier to uptake of filtering among PWID (Caflisch et al. 1999; Keijzer and Imbert 2011). As noted in Fig. 2, there was no evidence that active drug was lost during filtration (except for Suboxone®), although there are a number of caveats to this. First, cigarette filters absorb much of the solution they are placed in, so remain slightly wet after use. We have demonstrated that at this stage only approximately half to two-thirds of the active dose has been successfully passed through the filter (McLean et al. 2009). However, successive rinses with small volumes (1–2 mL) of sterile water improve recovery to 90–95% (McLean et al. 2009) (Fig. 2). Second, buprenorphine in buprenorphine/naloxone products produce a particular challenge for filtration, with recovery in our studies typically falling beneath 80%. This has been replicated in our laboratory for both tablets and sublingual strip in separate studies. Further work has examined possible contributions to this: despite centrifugation at 10,000 RPM for 15 min, extracts from 8 mg Suboxone® tablets dissolved in 3 and 4 mL of sterile water only recovered approximately half of the buprenorphine content. This is substantially lower than the expected solubility of buprenorphine in water (18.09 mg/mL at 37 °C and pH 4.10). However, both naloxone (TGA 2011) and several of the tablet excipients (e.g. lactose, mannitol, and maize starch) are more readily water soluble, which may interfere the dissolution of buprenorphine. As such, the insoluble portion of buprenorphine may have been filtered along with other insoluble excipients, contributing to the proportionate poorer recovery.

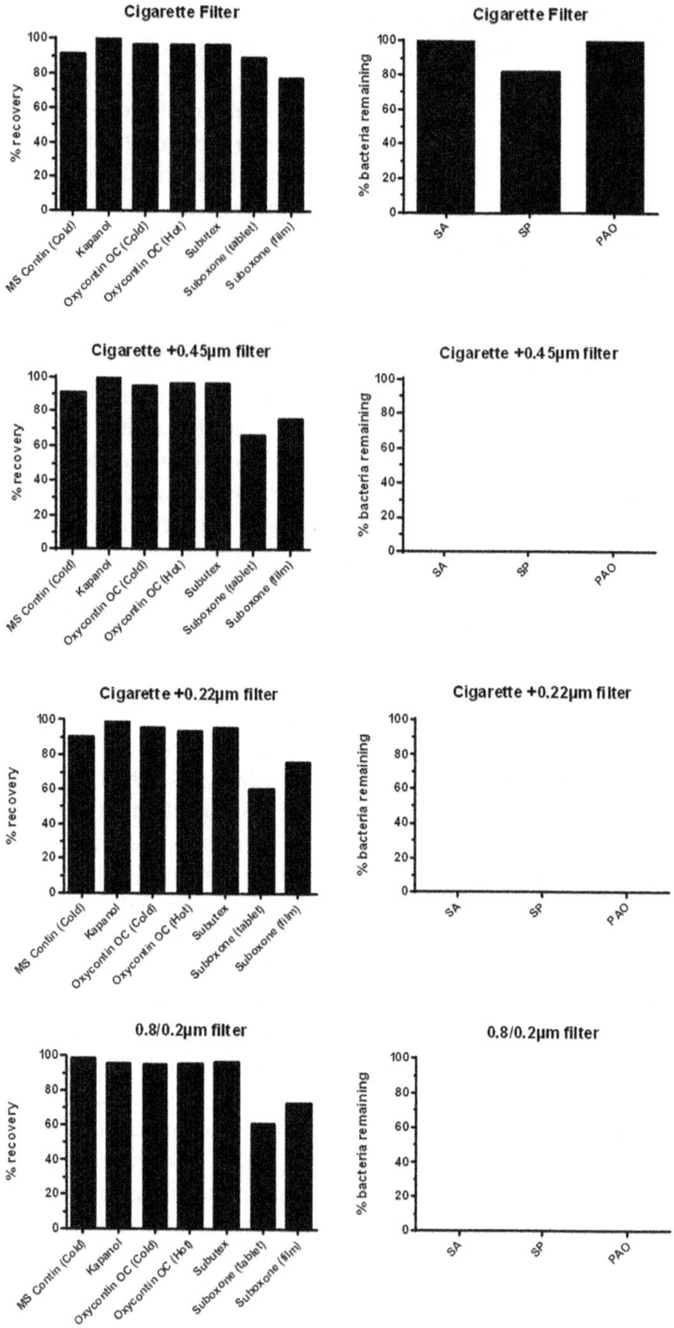

Fig. 2 Summarised results of studies of recovery of active opioid (*left*) and removal of common bacteria (*right*)

10 Summary and Conclusions

An optimal filter for use by PWID would possess a combination of high effectiveness at removal of both bacterial and particulate contamination; high recovery of active drug; simple and quick filtration procedure and low cost of device (Keijzer and Imbert 2011). Our studies here have demonstrated that the current devices fill many but not all of these criteria. Certainly, all of the devices incorporating an 0.2 μm filter membrane have demonstrated excellent effectiveness at removal of both particulate and bacterial contamination, albeit not to the standards of the British Pharmacopoeia. The single membrane device required manual pre-filtering with a coarse filter to avoid blockage, which increases the degree of handling and preparation required for injection, and the time required for the filtration process—both these issues are barriers to adoption of filtration, particularly in the case of individuals injecting in public or in other areas vulnerable to discovery (Keijzer and Imbert 2011). Indeed the time required to filter using these dual step processes may also prove a barrier where consumers are in withdrawal (Keijzer and Imbert 2011). The dual-membrane device (0.8/0.2 μm) was superior in terms of effectiveness, drug recovery, handling and time required for filtration. However, in contrast to cheaply available cigarette filters, these devices are relatively expensive (AUD ~ $2–4 each in contrast to a few cents for cigarette filters), and are not readily available.

While not yet optimised, the evidence from a range of tablet formulations demonstrates that the 0.2 μm filters provide a very substantial reduction of risks from particulate and bacterial contamination. Together with other safer injecting practices (handwashing, use of sterile water, alcohol swabbing at the injection site, and use of sterile injecting equipment), filtration would be expected to result in a far lower incidence of complications among PWID. A recent review of rates of bacterial infections among international studies of PWID suggested that the annual prevalence of symptoms is 6–36% (Hope 2010). The estimated cost to the Australian public health system of treating injection-related injury and disease in 2005/6 was $A25 million (range $21–35 m) (Sweeney et al. 2009), in a country with an estimated prevalence of PWID of 1.09% (range 0.65–1.5%), or approximately 150,000 (range 89,000–240,000) (Mathers et al. 2008). Even despite the cost of these filters, given these figures, it is clear that facilitating subsidised access to such an effective means of filtration through needle and syringe programs would be economically justifiable, as well as benefitting the health and well-being of people who inject drugs. Indeed, given the potential health benefits from filtration it is time to consider effective filters as essential a device as sterile needles/syringes for the world's approximately 16 million (range 11–21) PWID (Mathers et al. 2008).

Acknowledgements We are grateful to the individuals that have contributed to this programme of work, including Susan Brandon, Barbara de Graaff, Heng Guo, Roger Latham, Nathan Murray, Huei Ng, Pankaj Patel, and Troy Wanandy.

References

Arnett EN, Battle WE, Russo JV, Roberts WC (1976) Intravenous-injection of talc-containing drugs intended for oral use—cause of pulmonary granulomatosis and pulmonary-hypertension. Am J Med 60:711–718

Bassetti S, Battegay M (2004) *Staphylococcus aureus* infections in injection drug users: risk factors and prevention strategies. Infection 32:163–169

Boylan JC, Fites AL, Nail SL (1996) Parenteral products. In: Banker GS, Rhodes CT (eds) Modern pharmaceutics. Marcel Dekker, New York, pp 441–487

British Pharmacopoeia Commission (2005) British pharmacopoeia formula preparation: general monographs. The Stationery Office, London

Caflisch C, Wang J, Zbinden R (1999) The role of syringe filters in harm reduction among injection drug users. Am J Public Health 89:1252–1254

Cherubin CE, Sapira JD (1993) The medical complications of drug addiction and the medical assessment of the intravenous drug user: 25 years later. Ann Intern Med 119:1017–1028

Chiang W, Goldfrank L (1990) The medical complications of drug abuse. Med J Aust 152:83–88

Darke S, Ross J, Kaye S (2001) Physical injecting sites among injecting drug users in Sydney. Aust Drug Alcohol Depend 62:77–82

Darke S, Duflou J, Torok M (2015) The health consequences of injecting tablet preparations: foreign body pulmonary embolization and pulmonary hypertension among deceased injecting drug users. Addiction 110:1144–1151

Davis MA, Taube RA (1978) Pulmonary perfusion imaging—acute toxicity and safety factors as a function of particle size. J Nucl Med 19:1209–1213

Decuzzi P, Pasqualini R, Arap W, Ferrari M (2009) Intravascular delivery of particulate systems: does geometry really matter? Pharm Res 26:235–243

Del Giudice P (2004) Cutaneous complications of intravenous drug abuse. Br J Dermatol 150: 1–10

Del Giudice P, Vandenbos F, Boissy C, Cua E, Marion B, Bernard E et al (2005) Cutaneous complications of direct intra-arterial injections in drug addicts. Acta Derm Venereol 85: 451–452

Dettmeyer RB, Verhoff MA, Brueckel B, Walter D (2010) Widespread pulmonary granulomatosis following long time intravenous drug abuse—a case report. Forensic Sci Int 197:E27–E30

Dwyer R, Topp L, Maher L, Power R, Hellard M, Walsh N et al (2009) Prevalences and correlates of non-viral injecting-related injuries and diseases in a convenience sample of Australian injecting drug users. Drug Alcohol Depend 100:9–16

Ebright JR, Pieper B (2002) Skin and soft tissue infections in injection drug users. Infect Dis Clin North Am 16:697–712

Ganesan S, Felo J, Saldana M, Kalasinsky VF, Lewin-Smith MR, Tomashefski JF (2003) Embolized crospovidone (poly N-vinyl-2-pyrrolidone) in the lungs of intravenous drug users. Mod Pathol 16:286–292

Gordon RJ, Lowy FD (2005) Current concepts—bacterial infections in drug users. N Engl J Med 353:1945–1954

Gotway MB, Marder SR, Hanks DK, Leung JWT, Dawn SK, Gean AD et al (2002) Thoracic complications of illicit drug use: an organ system approach. Radiographics 22:S119–S135

Griffith CC, Raval JS, Nichols L (2012) Intravascular talcosis due to intravenous drug use is an underrecognized cause of pulmonary hypertension. Pulm Med 2012:617531

Haywood A, Glass BD (2011) Pharmaceutical excipients—where do we begin? Aust Prescriber 34:112–114

Hind CRK (1990) Pulmonary complications of intravenous drug misuse. 2. Infective and HIV related complications. Thorax 45:957–961

Hope V (2010) Neglected infections: real harms: a global scoping of injection-related bacterial infections and responses. In: Cook C (ed) The global state of harm reduction 2010 key issues for broadening the response. International Harm Reduction Association, London, pp 89–95

Illum L, Davis SS, Wilson CG, Thomas NW, Frier M, Hardy JG (1982) Blood clearance and organ disposition of intravenously administered colloidal partciles—the effects of particle size, nature and shape. Int J Pharm 12:135–146
Jorens PG, Van Marck E, Snoeckx A, Parizel PM (2009) Nonthrombotic pulmonary embolism. Eur Respir J 34:452–474
Kanke M, Simmons GH, Weiss DL, Bivins BA, Deluca PP (1980) Clearance of ^{14}Ce-labelled microspheres from blood and distribution in specific organs following intravenous and intraarterial administration in beagle dogs. J Pharm Sci 69:755–762
Kaushik KS, Kapila K, Praharaj AK (2011) Shooting up: the interface of microbial infections and drug abuse. J Med Microbiol 60:408–422
Keijzer L, Imbert E (2011) The filter of choice: filtration method preference among injecting drug users. Harm Reduction J 8
Kumar V, Abbas AK, Fausto N (2005) Robbins and Cotran pathological basis of disease. Elsevier Saunders, Philadelphia
Kuramoto K, Shoji T, Nakagawac Y (2006) Usefulness of the final filter of the IV infusion set in intravenous administration of drugs—contamination of injection preparations by insoluble microparticles and its causes. Yakugaku Zasshi-J Pharm Soc Jpn 126:289–295
Lim CCT, Lee SH, Wong Y-C, Hui F (2009) Embolic stroke associated with injection of buprenorphine tablets. Neurology 73:876–879
Low SU, Nicol A (2006) Talc induced pulmonary granulomatosis. J Clin Pathol 59:223
Mathers BM, Degenhardt L, Phillips B, Wiessing L, Hickman M, Strathdee SA et al (2008) Global epidemiology of injecting drug use and HIV among people who inject drugs: a systematic review. Lancet 372:1733–1745
McLean S, Bruno R, Brandon S, de Graaff B (2009) Effect of filtration on morphine and particle content of injections prepared from slow-release oral morphine tablets. Harm Reduction J 6
Miller DA, DiNunzio JC, Williams RO (2008) Advanced formulation design: improving drug therapies for the management of severe and chronic pain. Drug Dev Ind Pharm 34:117–133
Murphy EL, DeVita D, Liu H, Vittinghoff E, Leung P, Ciccarone DH et al (2001) Risk factors for skin and soft-tissue abscesses among injection drug users: a case-control study. Clin Infect Dis 33:35–40
Ng H, Patel RP, Bruno R, Latham R, Wanandy T, McLean S (2015) Filtration of crushed tablet suspensions has potential to reduce infection incidence in people who inject drugs. Drug Alcohol Rev 34:67–73
Orangio GR, Pitlick SD, Dellalatta P, Mandel LJ, Marino C, Guarneri JJ et al (1984) Soft-tissue infections in parenteral drug-abusers. Ann Surg 199:97–100
Ott MC, Khoor A, Scolapio JS, Leventhal JP (2003) Pulmonary microcrystalline cellulose deposition from intravenous injection of oral medication in a patient receiving parenteral nutrition. J Parenter Enteral Nutr 27:91–92
Pare JA, Fraser RG, Hogg JC, Howlett JG, Murphy SB (1979) Pulmonary 'mainline' granulomatosis: talcosis of intravenous methadone abuse. Medicine 58:229–239
Patel P, Patel RP, Brandon S, McLean S, Bruno R, de Graaff B (2012) Effects of filtration on the presence of particulate and oxycodone content of injections prepared from crushed OxyContin tablets. Curr Drug Saf 7:218–224
Roux P, Carrieri MP, Keijzer L, Dasgupta N (2011) Reducing harm from injecting pharmaceutical tablet or capsule material by injecting drug users. Drug Alcohol Rev 30:287–290
Rowe RC, Sheskey PJ, Cook WG, Quinn ME (2012) Handbook of pharmaceutical excipients. The Pharmaceutical Press, London, p 1033
Sameer HK (2010) Porosity measurements of cellulose acetate filters for some brands of cigarette using mercury porosimeter. Al-Mustansiriyah J Sci 21:57–66
Saydain G, Singh J, Dalal B, Yoo W, Levine DP (2010) Outcome of patients with injection drug use associated endocarditis admitted to an intensive care unit. J Crit Care 25:248–253
Schneider P, Duong TA, Ortonne N, Bagot M, Bouaziz JD (2010) Livedoid and necrotic skin lesions due to intra-arterial buprenorphine injections evidenced by maltese cross-shaped histologic bodies. Arch Dermatol 146:208–209

Schoenberger SD, Agarwal A (2013) Talc retinopathy. N Engl J Med 368:852
Scott J (2005) Laboratory study of the effectiveness of filters used by heroin injectors. J Subst Use 10:293–301
Scott J, Kennedy EJ, Winfield AJ, Bond C (1998) Investigation into the effectiveness of filters for use by intravenous drug users. Int J Drug Policy 9:181–186
Seeley RR, Stephens TD, Tate P (2006) Anatomy and physiology. McGraw Hill, New York
Shargel L, Wu-Pong S, Yu ABC (2005) Applied biopharmaceutics and pharmacokinetics. McGraw-Hill, New York
Sigdel S, Gemind JT, Tomashefski JF Jr (2011) The Movat pentachrome stain as a means of identifying microcrystalline cellulose among other particulates found in lung tissue. Arch Pathol Lab Med 135:249–254
Smith ME, Robinowitz N, Chaulk P, Johnson KE (2015) High rates of abscesses and chronic wounds in community-recruited injection drug users and associated risk factors. J Addict Med 9:87–93
Sousa C, Botelho C, Rodrigues D, Azeredo J, Oliveira R (2012) Infective endocarditis in intravenous drug abusers: an update. Eur J Clin Microbiol Infect Dis 31:2905–2910
Stafford J, Burns L, Trends Australian Drug (2014) Findings from the Illicit Drug Reporting System (IDRS). National Drug and Alcohol Research Centre, Australian Drug Trend Series Australia, p 2015
Sweeney R, Conroy AB, Dwyer R, Aitken CK (2009) The economic burden to the public health system of treating non-viral injecting-related injury and disease in Australia (a cost of illness analysis). Aust N Z J Public Health 33:352–357
TGA (2011) Australian Public Assessment Report for Buprenorphine/Naloxone. In: DOHA (ed). Canberra
Tomashefski JF Jr, Hirsch CS (1980) The pulmonary vascular lesions of intravenous drug abuse. Hum Pathol 11:133–145
Tung M (2015) Evolving epidemiology of injecting drug use-associated infective endocarditis: a regional centre experience. Drug Alcohol Rev 34:412–417

Misuse of Methylphenidate

David B. Clemow

Abstract This chapter reviews methylphenidate misuse, abuse, dependence, diversion, and malingering associated with its use as a prescription medication for attention-deficit/hyperactivity disorder and the nonmedical use linked to its stimulant effects. Methylphenidate-induced regional elevations in brain dopamine appear to be integral to both efficacy in attention-deficit/hyperactivity disorder and potential for abuse, raising potential concerns for drug safety and prescription drug diversion costs associated with nonmedical use. Regardless, methylphenidate is an important treatment option, and detecting malingering for the purpose of illicit access to methylphenidate for subsequent misuse or diversion is a difficult challenge. Also discussed are the effects of methylphenidate in patients with comorbid substance use disorder and the potential linkage of methylphenidate use with subsequent substance abuse. The current data suggest that methylphenidate misuse and diversion are common health-care problems with a stimulant prescription drug diversion prevalence of approximately 5–10 % of high school students and 5–35 % of college students. The effectiveness and speed of action of methylphenidate are deemed desirable to enhance attention and focus performance for activities such as studying for exams, but methylphenidate is also misused recreationally. These data suggest a need for close screening and therapeutic monitoring of methylphenidate use in the treatment of attention-deficit/hyperactivity disorder.

Keywords Methylphenidate · Misuse · Abuse · Dependence · Diversion · Malingering · ADHD

D.B. Clemow (✉)
Senior Clinical Research Scientist, Lilly Corporate Center, 46285 Indianapolis, IN, USA
e-mail: davidclemow@lilly.com

Curr Topics Behav Neurosci (2017) 34:99–124
DOI 10.1007/7854_2015_426
Published Online: 23 December 2017

Contents

1 Introduction

This chapter summarizes the current understanding of methylphenidate (MPH) misuse, abuse, dependence, diversion, and malingering in association with its use as a prescription medication in the treatment of attention-deficit/hyperactivity disorder (ADHD), which is not only a public health concern for the end users but is also a financial liability for insurance payers (Aldridge et al. 2011; Clemow and Walker 2014). This chapter pulls heavily from a recent comprehensive review on the potential for misuse and abuse of medications in ADHD (Clemow and Walker 2014) and expands upon MPH-specific data that is the focus of this chapter.

A licit medication, such as MPH, is a drug legally permissible under international control and used for medically prescribed purposes. Misuse or nonmedical medication use (NMU) occurs when a licit medication is used for a purpose for which it was not prescribed and when used to garner a nontherapeutic or nonmedical effect, such as for performance enhancement or recreational use. NMU is often associated with diversion and malingering. A drug is considered to be diverted when it is channeled (sold or given) into an illicit (illegal; nonauthorized) market for use generally not medically authorized or necessary. Illicit markets and desired nonmedical drug effects (e.g., stimulant effect) can lead to malingering, which occurs when a person falsely or grossly exaggerates a medical complaint with the goal of receiving a reward, such as a prescription for a medication that can be illegally marketed or used recreationally.

NMU can also be associated with abuse, which in turn can escalate to dependence. Because misuse and even abuse do not equate to dependence, and because the terms are often used interchangeably, understanding the differences among these terms in regard to this chapter is important. Abuse develops when a recreational user consumes the substance in amounts or with methods that are harmful to themselves or others. A cluster of cognitive, behavioral, and physiological symptoms then develops that indicate the continued use of the substance despite significant substance-related problems (American Psychiatric Association 2013). For example, agitation, irritability, and increased heart rate can be the symptoms of MPH abuse (National Institute of Drug Abuse 2011). In contrast to how defined in this chapter, misuse is often thought of as using the drug differently than prescribed but not for the intention of getting "high" (e.g., doubling the dose to gain efficacy), while abuse is often thought of as misuse with the intention of getting "high" (e.g., gain a euphoric feeling). Drug dependence can occur when the body develops a physical need for the drug, based on underlying central nervous system (CNS) changes that can persist after stopping the drug. The person develops cravings for and tolerance to the drug that can lead to greater use. Stopping the use of the drug can result in a specific withdrawal syndrome, thus prompting the individual to consume the substance to relieve the resulting symptoms (American Psychiatric Association 2013). Addiction is often considered a step beyond dependence, where a person has both a physical and a psychological need for a drug.

MPH misuse is an important topic within health care because MPH is a commonly prescribed medication for ADHD, which is a common disorder, and MPH is a psychostimulant classified as a schedule II drug by the Controlled Substances Act. By definition, schedule II drugs have a significant danger for abuse, and MPH's mechanism of action (MOA) and stimulant properties have been associated with misuse and abuse (Concerta Package Insert 2010), which can lead to safety issues (National Institute of Drug Abuse 2011; Schneider and Enenbach 2014). This is relevant not only as a public safety topic in general, but also a specific health care spend issue. Malingering and drug diversion include expenses in addition to the prescription costs, such as doctor visits, emergency department (ED) treatment, and other societal costs (Center for Medicare and Medicaid Services 2012).

1.1 Attention-Deficit/Hyperactivity Disorder (ADHD)

A diagnosis of ADHD is common in children (5–10 %) and often continues into adulthood (American Psychiatric Association 2013; Bloom et al. 2013). In the USA, the estimated prevalence of ADHD in adults is 4.4–5.2 % (Kessler et al. 2006; Fayyad et al. 2007). ADHD is a chronic, neurobiological behavioral disorder for which up to 60 % of children and adolescents with ADHD continue to have symptoms into adulthood (American Psychiatric Association 2013; Kessler et al. 2005). The relatively high adulthood prevalence is significant, as medication misuse and abuse are prominent in the young adult age group (Johnston et al. 2012). Moreover, despite continuing stigmatization of adult ADHD (Fuermaier et al. 2012), prescriptions for adult ADHD are increasing steadily, with prescriptions at least equal to the number of prescriptions for US children (Cascade et al. 2008).

Malingering for ADHD medications is difficult to discern, occurs relatively frequently, has increased over time, and occurred in the form of symptom exacerbation almost 50 % of the time in 1 studied college student ADHD population (Sullivan et al. 2007; Suhr et al. 2008; Musso and Gouvier 2014). Because malingering (faking symptoms) can be an issue in seeking stimulant medication prescriptions and because MPH's effects of increased attention and focus is core to its desire as a performance enhancer and recreational drug, it is important for physicians to recognize the 18 diagnostic symptoms of ADHD (Clemow and Walker 2014). People with ADHD are characterized by inattention and/or hyperactivity and impulsivity, primarily presenting with inattentiveness, hyperactivity/impulsivity, or a combination of both (American Psychiatric Association 2013). The inattentive symptoms are as follows: fails to give close attention to details or makes careless mistakes; difficulty sustaining attention; does not appear to listen; struggles to follow through on instructions; difficulty with organization; avoids or dislikes tasks that require sustained mental effort; loses things; easily distracted; and forgetful in daily activities. The hyperactive/impulsive symptoms are as follows: fidgets with hands or feet; difficulty remaining seated; feeling restless; difficulty engaging in activities quietly; feels like driven by a motor; talks excessively; cannot wait for turn in conversations; difficulty waiting or taking turns; and interrupts or intrudes upon others. For the inattentiveness and hyperactivity/impulsivity types of ADHD, an adult must exhibit ≥5 of 9 symptoms from the attention or hyperactive/impulsive category for at least 6 months. For combined type ADHD, they must present at least 5 of 9 symptoms from each category for 6 months.

1.2 Link Between ADHD Neurobiology, Medication Mechanism, and Abuse Potential

The neurobiology of ADHD and the MOA of the medications that treat it are critical to understanding the abuse potential of ADHD medications such as MPH (Clemow and Walker 2014). ADHD is a behavioral disorder believed to be caused by neurochemical signaling dysregulation in the brain, particularly the prefrontal cortex (PFC) that is involved in attention and executive functioning (Del Campo et al. 2011). Neuronal circuits in the PFC associated with inattention, hyperactivity, and impulsivity are linked to other areas of the brain associated with behavior; these brain regions include the basal forebrain, striatum, nucleus accumbens (NAc), thalamus, hypothalamus, amygdala, hippocampus, brainstem, and cerebellum (Stahl 2008). These neuronal projections regulate behavior such as movement, pleasure, and arousal via catecholamine neurotransmitter signaling that includes dopamine (DA) and norepinephrine (NE). Dysregulation in these catecholamine signaling pathways is the putative mechanism behind the diagnostic symptoms of ADHD (Stahl 2008; Arnsten and Rubia 2012).

MPHs, including dexmethylphenidate (d-MPH), are believed to block DA transporter (DAT) and NE transporter (NET) reuptake into the presynaptic neuron, thereby facilitating these neurotransmitter's signaling activities; MPH may also increase DA and NE direct release from the presynaptic terminal into the extra-neuronal space (Arnsten 2011). Medications such as MPH that increase DA and NE signaling can have positive effects on the symptoms of ADHD (Arnsten and Li 2005; Arnsten 2011; Murnane and Howell 2011; Arnsten and Rubia 2012).

The neural basis of MPH's potential for drug abuse is likely based upon its effects on DA signaling within the amygdala, NAc, ventral striatum, and possibly the orbital PFC (Whitelaw et al. 1996; Hutcheson and Everitt 2003; Ito et al. 2004). DA-dependent mechanisms in these brain structures mediate the reinforcing and stimulant effects of drugs of abuse and play a major role in abuse behavior (Ito et al. 2000, 2004; Rowlett et al. 2007; Economidou et al. 2011). Medications such as MPH that affect DA within these brain regions can be at risk for abuse potential. In fact, abuse and even dependence have been observed relatively frequently with stimulant medication misuse; approximately 13 % of stimulant medication misusers in the previous year met the 2004 National Survey on Drug Use and Health (NSDUH) criteria for abuse or dependence (Kroutil et al. 2006).

There are multiple medications other than the stimulants MPH and d-MPH that are approved for the treatment of ADHD, with some having abuse potential and others not, based upon their MOA (Clemow and Walker 2014). Stimulants in addition to MPH include the amphetamines (AMPs): dextroamphetamine (d-AMP), methamphetamine, lisdexamfetamine, and mixed AMP salts. The AMPs are believed to work similarly to MPH by blocking DAT and NET reuptake (Arnsten 2011). AMP stimulants also increase the levels of DA in the NAc and other areas of the brain associated with stimulant effects, euphoric effects, and drug abuse. As with MPH, the AMP stimulants have been associated with abuse potential (Clemow and Walker 2014). Although the effects of these stimulants may be less pronounced in the NAc than the PFC, explaining why they do not appear to cause addiction when used as prescribed (Arnsten and Rubia 2012) and why actual MPH abuse may be limited in real-world settings (Kollins 2007). With that said, the CNS stimulant effect of the MPHs and AMPs in cases of consistent misuse could potentially lead to abuse in some cases.

There are also nonstimulants approved for the treatment of ADHD, which are the α_2-adrenergic receptor agonists (guanfacine and clonidine) and atomoxetine (ATX). Their MOAs are different than that of the stimulants, which is of note since the nonstimulants do not appear to have potential for abuse, dependence, or withdrawal (Heil et al. 2002; Arnsten and Li 2005; Jasinski et al. 2008; Kawaura et al. 2014). The α_2-adrenergic receptor agonists directly mimic the effects of NE at postsynaptic α_{2A}-receptors (Arnsten 2011; Arnsten and Rubia 2012) and do not appear to influence DA in CNS areas linked to abuse (Arnsten and Li 2005). The MOA of ATX appears to be as a selective inhibitor of the presynaptic NET (Bymaster et al. 2002), thereby increasing α_2-receptor activation via elevated synaptic NE (Gamo et al. 2010). Interestingly, although ATX is selective for NETs, ATX appears to increase DA levels selectively (regionally specific) in the PFC (Stahl 2003a, c), most likely through nonspecific action of the NET (Arnsten 2011). Despite this finding, ATX does not have any effect on the DAT in the NAc or striatum (Stahl 2003b), which may explain why ATX does not appear to be associated with abuse potential (Bymaster et al. 2002). The lack of effect on DA in reward centers of the brain and the lack of feeling a perceptible arousal, euphoric, or liking effect that can occur with stimulants likely underlie the nonstimulants lack of abuse or dependence, as well as their very limited misuse or diversion (Heil et al. 2002; Simpson and Plosker 2004; Jasinski et al. 2008; Clemow and Walker 2014). Guanfacine, clonidine, and ATX are unscheduled and unrestricted medications (Kapvay Package Insert 2010; Intuniv Package Insert 2011; Strattera Package Insert 2014).

1.3 Prevalence of ADHD Stimulant Medication Misuse

A systematic review of 21 studies indicated that the prevalence of stimulant (MPHs and AMPs) misuse within the general population during the previous year ranged from 5 to 9 % in school-age children (all grades) and 5–35 % in college-age students (Wilens et al. 2008b). Multiple studies have found rates of misuse within these ranges that vary somewhat based upon location or specific college included in the study (McCabe et al. 2005; Rabiner et al. 2009a, b; Rabiner 2013; McCabe and West 2013). The NMU of stimulants appears to be equally common among men and women, while higher rates exist in Caucasians and Hispanics compared with other nationalities and fraternity and sorority members compared with nonmembers (Merkel and Kuchibhatla 2009). Stimulant NMU was found to be >3 times higher in Caucasians and Hispanics than African American college students (Teter et al. 2006). Other factors contributing to an increased prevalence of NMU include young adult age, other illicit substance use, mental health comorbidities, and SUD comorbidities (Merkel and Kuchibhatla 2009; Chen et al. 2014).

In one study of 1536 middle and high school students in the USA, NMU of stimulant medication was reported by 4.5 % of the study population. Of the students who reported prescription stimulant use, 23.3 % reported being approached to sell, give, or trade their prescription drugs (McCabe et al. 2004a). The NMU of stimulant ADHD medications by students is equally driven by wanting to be more productive and/or stay awake (performance enhancement) and wanting to feel good or get high (recreational) (Upadhyaya et al. 2010). Thus, NMU of stimulant medications is not uncommon, and approximately one-fourth of patients with ADHD have been approached by others who want to purchase their stimulant medication (McCabe et al. 2004a). Misuse of stimulant ADHD medications is more prevalent in patients with ADHD and comorbid substance use disorder (SUD), which is relevant because almost 30 % of adult patients with ADHD are thought to have comorbid alcohol and drug dependency (Biederman 2005). Moreover, the rate of life-time drug abuse is 4 times greater in adults with ADHD than those without (Upadhyaya 2007).

In concert with NMU, a large proportion of patients divert ADHD stimulant medications from their prescriptions, with 14.7 % of students in grades 7, 9, 10, and 12 reporting they gave away their medications and 7.3 % reporting they sold their medications (Poulin 2001). In college students prescribed medications for ADHD, almost 29 % reported giving or selling their medication to someone else (Upadhyaya et al. 2005). In an adult ADHD population covered by private health insurance, 22.5 % of men and 9.1 % of women (16.6 % overall) diverted their medication from their prescriptions (Aldridge et al. 2011).

Diversion for NME is linked to malingering to obtain a diagnosis of ADHD in order to obtain stimulant medications, and this practice is apparently not uncommon. As many as 50 % of university students who self-referred for an ADHD evaluation appeared to have exaggerated their symptoms in a noncredible manner in 2 different studies (Sullivan et al. 2007; Suhr et al. 2008). In a nationally representative study of adults, nearly 20 % of nonmedical users from the previous year indicated that they had misrepresented their symptoms to obtain a prescription (Novak et al. 2007).

Based upon 2008 data from the study of ADHD patients covered by private health insurance, after controlling for copayments and other variables, the estimated value of diverted MPH in a 30 day period was about $3 million US dollars, representing about 4.3 % of the total costs that private insurers paid for MPH medications (Aldridge et al. 2011). Of note, the relatively high incidence of MPH diversion in this study appeared to be driven by formulations other than Concerta® (extended-release form of MPH) which had only a 0.2 % of insurer cost diverted. Long-acting forms of stimulants have been developed purposefully to reduce the fast onset of euphoria experienced with short-acting forms in hopes to reduce abuse potential, which has been successful (Jasinski and Krishnan 2009). Payers, such as managed care and Medicaid, may need to be concerned about stimulant diversion for NMU due to wasted pharmacy spend on diverted drugs and subsequent follow-up treatment costs of stimulant abusers who are seen in the ED (Aldridge et al. 2011; Substance Abuse and Mental Health Services Administration 2013).

More importantly, stimulant misuse can be associated with clinically significant safety outcomes, as is demonstrated by the number of ED visits associated with stimulant misuse noted by the Drug Abuse Warning Network, a public health surveillance system that monitors drug-related ED visits in the USA. Between 2005 and 2010, ED visits associated with ADHD stimulant medications (a significant portion of which were MPH) increased from 13,379 to 31,244, and visits resulting from the NMU of stimulant drugs nearly tripled from 5212 to 15,585 (Substance Abuse and Mental Health Services Administration 2013). The 18–25 year-old age bracket had the highest incidence of ED visits linked to stimulant misuse. In one study of stimulant medication linked ED visits, 37 % of cases involved stimulant medications exclusively, while the remainder of cases included other drugs, such as other pharmaceuticals and alcohol (Rabiner 2013).

These data support the importance of understanding the misuse of stimulant medications prescribed for ADHD. This chapter focuses on MPH because it is such a widely prescribed medication for ADHD; for example, in a US managed care population, about 40 % of ADHD prescriptions were for some formulation of MPH (Christensen et al. 2010).

2 Methylphenidate Misuse

MPH NMU is a microcosm of stimulant medication (MPHs and AMPs) misuse in general, with MPH-specific data generally mimicking results of studies looking at stimulant medication misuse in general. Thus, for example, while there are limited data on malingering specifically for MPH, studies of malingering for stimulants in general are likely generally applicable.

2.1 *Methylphenidate Abuse Liability*

MPHs facilitate DA neurotransmission in the NAc and striatum subcortical areas of the CNS, which is thought to contribute to the psychomimetic effects, euphoria, and abuse potential that can be associated with MPH use (Kuczenski and Segal 1997; Bymaster et al. 2002; Bredeloux et al. 2007; Koda et al. 2010; Stahl 2003b). The fact that DA is being increased in these areas via DAT blockade is pertinent since data supports the notion that the reinforcing effects of cocaine may also be mediated through a similar mechanism (Volkow et al. 1997). Additionally, MPH appears to cause the release of DA through the DAT by 2 mechanisms that include a slow cytosolic DA exchange and a fast direct DAT channel mechanism (Kahlig et al. 2005). This fast mechanism may be what leads to stimulant reinforcing effects (Volkow and Swanson 2003).

A multitude of preclinical studies has been conducted to elucidate the nuance of the neurochemical mechanisms and abuse behaviors linked to the possible abuse potential of MPH. In a self-administration lever-press reward study, rhesus monkeys were trained to press one lever for injections of either saline or drug and another lever for food (Gasior et al. 2005). In this study comparing abuse liability in monkeys dosed with behaviorally active levels of cocaine, MPH, d-AMP, ATX, and desipramine, those monkeys dosed with cocaine, MPH, and d-AMP had an increased lever-press response of >90 % while those dosed with ATX and desipramine did not maintain self-administration behavior above vehicle levels. These results were supported in another monkey study using a different model believed to be predictive of abuse potential in humans (Wee and Woolverton 2004). In this second study, monkeys were prepared with chronic intravenous catheters and trained to press a lever to receive cocaine injections. A drug dose was considered a positive reinforcer in any monkey if the number of injections it maintained in all test sessions was outside the 95 % confidence interval for all vehicle test sessions. MPH and d-AMP dosed monkeys showed high levels of repeated self-administration, whereas ATX and desipramine did not. MPH consistently maintained responding above the levels maintained by its administration vehicle.

Work in rats suggests MPH treatment in adolescents can alter responsiveness to cocaine during adulthood through enduring neuronal changes in brain reward system neurobiology. Whether adolescent MPH treatment increased or decreased sensitivity to cocaine reward (abuse/addiction liability) in those rats as adults was dependent upon the age of the rat when MPH exposure occurred (Andersen et al. 2002; Schenk and Izenwasser 2002; Brandon et al. 2003; Carlezon et al. 2003). In contrast to adolescent rats, adult rats were more susceptible to MPH exposure leading to greater abuse liability.

Complicating matters, other studies have shown that adolescent mammals are more susceptible to the effects of reward while less susceptible to the aversive drug effects compared with adults (Schramm-Sapyta et al. 2007; Doremus-Fitzwater et al. 2010). In a study of adolescent and adult rats, there was a blunted expression of MPH as measured by a conditioned taste avoidance test in the younger cohort, suggesting that adolescents may be protected from the aversive effects of MPH compared with adults (Wetzell et al. 2014). This could contribute to younger rats being more likely to abuse MPH than adult rats. This may translate to humans where studies of MPH NMU show that NMU occurs primarily in high school-age and college-age students and is less common in those in their late 20 s and older (Bogle and Smith 2009).

Several human studies have been conducted examining MPH abuse liability (Clemow and Walker 2014). In a randomized, double-blind, placebo (PBO)- and comparator-controlled (90 mg ATX) study, the abuse liability of MPH (40 mg) was examined in 16 healthy adults who were experienced recreational drug users (Heil et al. 2002). Study participants answered questions about their experienced drug effects up to 240 min after dosing based on a visual analog scale (VAS) of "liking," which is an empirically derived drug Addiction Research Center Inventory (ARCI) tool. An adjective rating scale (ARS) was also employed. MPH produced increases in many self-report measures sensitive to stimulant effects, including the stimulant scales of the ARS, VAS, and the ARCI benzedrine, AMP, and morphine-benzedrine subscales (Heil et al. 2002). In contrast, participants receiving ATX did not report positive drug effects different from PBO.

In a randomized, double-blind, PBO- and comparator-controlled, single-dose cross-over study conducted in 40 stimulant-preferring drug abusers, subjects received either 90 mg MPH, 60 mg phentermine, 100 or 200 mg desipramine, 45, 90, or 180 mg ATX, or PBO (Jasinski et al. 2008). Phentermine was used as a positive control; desipramine and PBO were used as negative controls, while the nonstimulant ATX was a study focus. Over a 24 h period after drug administration, a Drug Rating Questionnaire and Drug Identification Questionnaire were used to assess euphoria. MPH was liked significantly more than any dosage of ATX, desipramine, or PBO. Both MPH and phentermine significantly increased euphoria scores compared with ATX, desipramine, and PBO (Jasinski et al. 2008).

Similar results were observed in a small study of 6 subjects with recent histories of stimulant use, where subjective effects of 5–30 mg MPH were compared with

15–90 mg ATX, 2.5–15 mg d-AMP, 0.06–0.375 mg triazolam, and PBO in humans who had learned to discriminate 30 mg MPH (Lile et al. 2006). MPH and d-AMP increased drug-appropriate responding and produced stimulant-like subjective effects that were significantly greater than PBO on multiple testing domains. Triazolam, a sedative-hypnotic used as a negative control, produced insignificant levels of drug-appropriate response. ATX did not produce effects significantly different from PBO.

While MPH has abuse potential, its neuropharmacologic and pharmacokinetic properties reduce its abuse potential considerably compared to other stimulant drugs of abuse, such as cocaine (Kollins 2003). Actual MPH abuse (in contrast to misuse) may be limited (Kollins 2007). In one small study (n = 33) comparing MPH to PBO, MPH produced reinforcing effects only for the ADHD group (n = 16) and not for the control group (n = 17) (Kollins et al. 2009); the reinforcing effects in the ADHD cohort were likely attributable to ADHD symptom control rather than euphoria-producing effects. MPH increased stimulant-related subjective effects observed in patients without ADHD that were not associated with drug reinforcement. However, MPH has been found to increase drug reinforcement as an increasing function of dose under performance conditions in a clinical research setting (Stoops et al. 2005); reinforcing effects of MPH appear to be sensitive to manipulation of behavioral demands (Rush et al. 2001; Stoops et al. 2004). Regardless, these types of data reiterate the importance of distinguishing between abuse and misuse (Kollins 2007), particularly when considering implications in real-world settings.

2.2 *Methylphenidate Diversion and Misuse*

While the level of MPH abuse may be controversial, based on a number of studies conducted through surveys and interviews incorporating responses from thousands of subjects, data suggest that MPH is misused frequently, particularly with high school and university students, who often use MPH to improve their concentration and alertness during long-term study sessions (performance enhancement) (McCabe et al. 2004b, 2005, 2006a, b; Teter et al. 2005). Remarkably, students who reported NMU of prescription stimulants exceeded those who reported medical use (McCabe et al. 2006b). The students misusing MPH were more likely to use other drugs, report lower grade point averages, and not attend college (McCabe et al. 2004a, b). In a survey of 2087 college students, 110 (5.3 %) admitted to MPH NMU at least once, and that they obtained the MPH through friends or family (Dupont et al. 2008). In an interview study of 50 students reporting MPH misuse, 70 % reported recreational use while 30 % reported use of MPH to aid in studying (Barrett et al. 2005). In a survey completed by 2250 college students, 3 % reported using MPH illicitly in the previous year (Teter et al. 2003). During the previous year of college as well as before attending college, the respondents reporting illicit MPH use were significantly more likely than those not using MPH to also use alcohol and other drugs. In a follow-up study of >4500 college students, nearly 6 % of respondents reported NMU of a stimulant within the previous year, of which one-fourth had taken MPH (Teter et al. 2006). Despite these observations, MPH NMU may be in a decline. The Monitoring the Future survey of drug use has shown that NMU of MPH in 8th, 10th, and 12th graders declined from 4 % of students in 2001 to 2 % of students in 2013 (Johnston et al. 2014).

In a survey of 66 adults who were prescribed MPH, 44 % admitted to diverting their medication and 29 % misused their medication (Darredeau et al. 2007). Of those who diverted their MPH, 97 % gave it away rather than selling it. On average, those diverting their medication were younger, had started their prescription at a younger age, and were more apt to use other illicit substances than those who did not divert their medication. Those ADHD patients who misused their MPH were more likely to have a comorbid SUD than those that did not misuse their medication. In another study, about 26 % of students with prescribed MPH gave or sold some of their medication (Poulin 2007). In a structured interview study of 55 ADHD subjects, 11 % reported selling their medication and an additional 22 % misused their medication; the ADHD subjects with conduct disorder or SUD were the ones diverting or misusing their medication, suggesting that monitoring medication use in those ADHD patients with these comorbidities is particularly relevant (Wilens et al. 2006). Additionally, high-risk individuals need to be monitored for the use of nonprescribed stimulants, and patients with ADHD need to be educated about the issues associated with misuse and diversion of their stimulant medications (Wilens et al. 2008b).

Because MPH is difficult to synthesize, direct diversion of prescribed MPH is the main source for NMU (Merkel and Kuchibhatla 2009). One way to reduce the NMU of MPH is to make preparations of the drug that make it difficult to crush into a fine powder for nasal inhalation or dissolve into a liquid for injection. These are methods of intake that can be used for abused psychostimulants that allow the drug to be more rapidly taken into the brain to produce euphoric effects. A study in recreational drug users has shown that immediate-release MPH has significantly greater subjective effects than extended-release MPH formulations (Parasrampuria et al. 2007). Comparing MPH immediate-release (generally short-acting) versus extended-release (generally long-acting) formulations using positron emission tomography has shown that higher and earlier peak plasma concentrations from the immediate-release form result in a quicker absorption of MPH into the brain (Spencer et al. 2006). This study also found that the extended-release formulation did not have the same level of likeability that was observed with the immediate-release formulation. These findings are consistent with results from other studies demonstrating that diversion and misuse are much more common for immediate-release than extended-release stimulant formulations (Cortese et al. 2013; Harstad et al. 2014); MPH misusers seek the more immediate and often more likeable effects of immediate-release formulations when used recreationally or for studying (Spencer et al. 2006; Bright 2008; Harstad et al. 2014), suggesting greater monitoring of patients seeking or using short-acting MPH formulations is needed.

2.3 Methylphenidate Misuse Safety Concerns

Intoxication effects and potential health risks that can be associated with MPH abuse include feelings of exhilaration, increased energy, mental alertness, increased heart rate, increased or decreased blood pressure, increased metabolism, digestive problems, loss of appetite, weight loss, nervousness, insomnia, and perhaps even seizures, heart attack, and stroke (National Institute of Drug Abuse 2011). Even if not abusing MPH in regard to quantity or frequency, someone taking nonprescribed MPH is likely to be unaware of its safety profile. MPH can increase systolic blood pressure (3–8 mm Hg), diastolic blood pressure (2–14 mm Hg), and pulse (3–10 beats/min), although epidemiological studies have shown stimulant use is not associated with cardiovascular symptoms, events, or sudden death (Cooper et al. 2011; Habel et al. 2011; Olfson et al. 2012; Schneider and Enenbach 2014). Child growth inhibition has been a historic concern, but it is debated, and evidence suggests that stimulant use is not associated with differences in growth over a 10- to 11-year follow-up (Biederman et al. 2010), and in cases of observed growth deficit, it tends to attenuate after stimulant discontinuation (Faraone et al. 2008). Tic development or worsening has been linked to MPH use, but the data are controversial, contrasting, and confounded by the fact that tics are a common comorbidity with ADHD (Schneider and Enenbach 2014). Thus, while cardiovascular events, growth inhibition, or tics generally may not be of major concern, there are common adverse events associated with MPH use that include abdominal pain, decreased appetite, headache, dry mouth, nausea, insomnia, anxiety, dizziness, decreased weight, irritability, and hyperhidrosis (Concerta Package Insert 2010); patients should be made aware of these potential effects.

Overall, the misuse of MPH raises legitimate safety concerns for overdose and drug interactions with other medications or NMU drugs, particularly since illicit users are generally unaware of these issues and often use MPH with other recreational drugs. Researchers found in one study that approximately 4 out of every 5 college students using ADHD prescription medications for NMU also reported heavy episodic drinking in the previous 2 weeks and were significantly more likely to use other NMU drugs (McCabe et al. 2006a). Stimulant medication misusers have been found to be significantly more likely to binge drink, use cocaine, and test positively on drug abuse screening tests (Sepúlveda et al. 2011).

In a study from the National Electronic Injury Surveillance System–Cooperative Adverse Drug Event Surveillance Project, excluding self-harm and drug abuse visits, of 188 ED visits linked to stimulant medication use, unintentional ingestion or overdose accounted for 61 % (Cohen et al. 2006), and 36 % of the cases were patients who were not prescribed the stimulant medication. Moreover, 14 % of the ED visits were associated with cardiovascular events. In an American Association of Poison Control Centers Toxic Exposure Surveillance System study from 1993–1999 of 759 patients aged 10 through 19 years with MPH toxic exposure, 42.7 % were in 10–14 year-old children (Klein-Schwartz and McGrath 2003). For 570 patients (75.1 %) managed in a health-care facility, 398 patients were discharged from the ED and 172 were admitted. Of the 530 patients (70.0 %) involving MPH only, the frequency increased sevenfold from 1993 to 1999. The most common symptoms in adolescents with MPH only exposure were tachycardia (31.7 %), agitation/irritability (25.7 %), and hypertension (11.5 %). The study reported that there were no symptoms in 24.9 % of the cases, while there were mild, moderate, and severe effects in 41.9, 32.3, and 0.9 % of patients, respectively (Klein-Schwartz and McGrath 2003).

In studies of stimulant medication misuse, adverse events were frequently reported, including sleep difficulties (72 %), irritability (62 %), dizziness and lightheadedness (35 %), headaches (33 %), stomachaches (33 %), and sadness (25 %), although most students also reported an overall positive effect (Rabiner 2013). As with what can occur with alcohol misuse, young adults seem to favor what they perceive as positive effects from the drugs over any deleterious effects they may have. A noteworthy finding was that approximately 5 % of those using ADHD medications for NMU believed that it contributed to use of other prescription drugs and illicit substances (Rabiner 2013).

In 2006, the Drug Abuse Warning Network estimated that 535,000 persons received substance abuse treatment for stimulants (Substance Abuse and Mental Health Services Administration 2007). Meanwhile, a substantial increase globally in the prescription rates of MPH has occurred during the last decade (Dalsgaard et al. 2014), and the rate of adult ADHD diagnosis and treatment continues to increase (Cascade et al. 2008). Approximately 7 % of adults in the USA have used a stimulant ADHD medication for NMU at least once (Novak et al. 2007).

3 Discussion

Taken together, the findings presented in this chapter suggest that MPH is misused and diverted frequently and has a pertinent level of abuse liability. While there does not appear to be a clear association between MPH use and long-term dependence or development of SUD, there is substantial evidence of MPH NMU for performance enhancement and recreation. Of note, college students prescribed stimulant medication in the previous year were more likely to be approached to divert their medication (54 %) than those prescribed pain medication (26 %), sedatives or anxiolytics (19 %), or sleeping medication (14 %) (McCabe et al. 2006a; Messina et al. 2014). In fact, the NMU to medical use ratio for stimulant medication was the highest among all 4 classes of prescription drugs. Among stimulant formulations, immediate-release stimulants are at a much greater risk for misuse than extended-release formulations, because patients favor the effects that immediate-release stimulants provide (Spencer et al. 2006; Bright 2008; Harstad et al. 2014).

Because MPH is a frequent treatment for ADHD, which is a common disorder, and because NMU of MPH medications is prevalent, NMU of MPH can be considered a public health concern. The NMU of MPH and other ADHD medications is likely to increase, as it has been shown that changes in medical use of prescription medications appear to mirror similar changes in the diversion and NMU of those same prescription medications among college students (McCabe et al. 2014). Over the last decade, past-year medical use of prescription stimulants for the treatment of ADHD rose from 1.9 % in 2003 to 4.7 % in 2013, paralleling an increase in past-year diversion and NMU from 5.4 % in 2003 to 9.3 % in 2013. NMU use of MPH is clinically relevant due to the multiple adverse effects that have been linked to NMU of MPH, particularly when used in combination with alcohol and other recreational drugs. MPH can be associated with withdrawal effects such as jitteriness and weight loss (Substance Abuse and Mental Health Services Administration 2001) as well as rebound effects that lead to ADHD symptoms worsening when the MPH medication wears off (Carlson and Kelly 2003).

While not a focus of this chapter, it is important to note that AMPs in recent years have been used more often than MPH for NMU, particularly AMP immediate-release formulations (Mao et al. 2011). In a survey of students reporting stimulant NMU in the previous year (5.9 %), more students (75.8 %) reported the use of AMP misuse compared with MPH misuse (24.5 %), suggesting that college students tend to misuse AMPs more than MPH (Teter et al. 2006). Several other studies support the finding that AMPs are misused more frequently than MPHs (Harstad et al. 2014; Berman et al. 2009). Immediate-release or short-acting formulations have been shown to be misused or abused most frequently (Harstad et al. 2014). The lower frequency of the abuse of MPH compared with amphetamines might be due to a lack of availability of intravenous or inhaled forms of MPH that provide fast delivery of the drug to the brain that seems to be needed to produce the pleasurable sensations often attributed to stimulants (Berman et al. 2009).

Due to the absent abuse liability, negligible misuse, and noncontrolled substance status of the nonstimulants guanfacine, clonidine, and ATX, they may be viable alternatives to MPH or AMP for the treatment of ADHD when patients do not respond to stimulants with optimal efficacy or have concerns regarding misuse, abuse, or SUD. The lack of misuse among nonstimulants is understandable since they appear to lack a quick onset of action desired for performance enhancement or a stimulant effect desired for recreational use (Clemow and Walker 2014).

In addition to issues of MPH general misuse and related safety concerns, malingering to obtain MPH for illicit use and issues with the cost of diversion are also concerns. Diversion for NMU costs payers millions of dollars every year, which is likely to increase as adult ADHD prescriptions continue to increase and young adults continue to misuse MPH. However, regardless of these potential troubles, MPH is a successful, well-studied option for the treatment of ADHD, and any misuse risk should be considered within this context, with a focus on awareness rather than avoidance.

3.1 Malingering

Incentives for malingering ADHD symptoms focus on gaining prescriptions for stimulant medications to enhance performance, but also include gaining additional school services and accommodations that are supported by the Americans with Disabilities Act (Sullivan et al. 2007; Sansone and Sansone 2011; Musso and Gouvier 2014); such accommodations include separate testing environments, longer testing times, reduced homework, and provision of a note taker. Another incentive is to gain stimulants for use recreationally or to illicitly sell the drugs for profit (Musso and Gouvier 2014).

Shopping for stimulant ADHD medications has been shown in a pharmacy database cohort study looking at more than 4 million subjects dispensed ADHD medications (Cepeda et al. 2014). Assessing the number of subjects with overlapping dispensings from different prescribers over 18 months from 2011 to 2012 showed that prescriptions from two or more prescribers dispensed by two or more pharmacies were most common in adolescents and young adults, occurring with about 200,000 subjects (4.5 % of subjects prescribed ADHD medication). The incidence drops to 0.27 % when dispensing was from two or more prescribers at three or more pharmacies, which was the definition of shopping behavior used by the authors.

Malingering has been shown to be difficult to detect in multiple studies (Rabiner 2013). Malingerers can easily manipulate the multitude of available ADHD diagnostic assessments and symptom questionnaires (Quinn 2003; Jachimowicz and Geiselman 2004; Harrison et al. 2007; Tucha et al. 2009; Lee et al. 2011; Sollman et al. 2010; Harp et al. 2011; Young and Gross 2011). While failure of additionally provided neuropsychological tests may have some usefulness in weeding out malingerers, findings to date suggest that differences between those with ADHD and malingerers are not distinct enough to detect feigned ADHD (Musso and Gouvier 2014; Musso et al. 2014). There is an unmet need for the development of assessments specifically designed to detect ADHD malingering. Encouraging work has been done examining ADHD malingerers tendency to more strongly feign hyperactivity and restlessness (Harrison et al. 2007) and intentionally perform poorly on cognitive and symptom validity tests (Marshall et al. 2010), suggesting that symptom exaggeration identification may be possible. Personality inventories, neuropsychological tests, and symptom validity scales may be useful in detecting feigned ADHD (Musso and Gouvier 2014; Sullivan et al. 2007; Jasinski et al. 2011; Quinn 2003; Sollman et al. 2010; Young and Gross 2011; Suhr et al. 2011). Unfortunately, results to date are variable, and these tools are likely unrealistic for use in real-world clinical settings. These findings support the need of health-care providers (HCPs) to receive and utilize medical information on this topic and critically assess patient needs, through direct observation as well as reports by others (e.g., parents), when diagnosing ADHD and assessing patient interest in secondary gain (Booksh et al. 2010; Joffe 2014).

3.2 *Methylphenidate Use and Subsequent Substance Abuse Disorder*

Being diagnosed with ADHD is associated with an elevated likelihood of also using nicotine, alcohol, and marijuana and other illicit drugs, regardless of age (Upadhyaya et al. 2005; Arias et al. 2008; Lee et al. 2011; Wilens et al. 2011; Wilens and Morrison 2012). Moreover, multiple studies have shown a linkage between ADHD and an increased risk of SUD (Biederman et al. 1995, 1997; Molina and Pelham 2003; Weiss and Murray 2003; Upadhyaya et al. 2005; Charach et al. 2011), with a high comorbidity rate of SUD in patients with ADHD (Cortese et al. 2013). Human positron emission tomography studies have shown reductions in DA synaptic markers in ADHD study participants that support a biological basis that may make patients with ADHD susceptible to SUD (Volkow et al. 2007a, b, 2009; Chen et al. 2014).

However, there is inconsistent data and opinion on whether the SUD that is often comorbid in patients with ADHD is solely linked to their ADHD neurobiology and ADHD symptomology, or if the common ADHD stimulant medication treatments for these patients may also be playing a role. In contrast to what one might hypothesize, studies have shown that use of MPH in patients with ADHD can actually reduce the likelihood of a later SUD by as much as twofold (Biederman et al. 1999; Wilens et al. 2003, 2008a; Katusic et al. 2005; Merkel and Kuchibhatla 2009; Groenman et al. 2013), and may even reduce the risk of SUD relapse (Wilson and Levin 2005). A study of children and adolescents with ADHD who were followed-up 10 years later suggested that stimulants neither increased nor decreased the risk for later SUD (Biederman et al. 2008).

In one study, the younger the age of a child when receiving an ADHD prescription for MPH, the less likely he or she was of having a subsequent SUD (Mannuzza et al. 2008). In a long-term study of 208 children and adolescents with ADHD, and a mean age of 31 years at time of follow-up showed that age at time of initiation of MPH treatment was correlated with later SUD potential. For every year the child was older at initiation of stimulants, the risk of SUD in adulthood increased by a factor of 1.46 (Dalsgaard et al. 2014), suggesting that stimulant effects on long-term SUD development may be age dependent. These human findings are consistent with some preclinical rat data showing that MPH treatment in adolescents can decrease cocaine responsiveness during adulthood (Andersen et al. 2002; Carlezon et al. 2003); however, other rat data, particularly in older rats, suggest that exposure to MPH can increase abuse liability in an age dependent manner (Schenk and Izenwasser 2002; Brandon et al. 2003). The potential for stimulant medication effects on development of SUD has also been examined in the spontaneously hypertensive rat (SHR) animal model of ADHD, with data supporting a potential MPH effect. In this SHR model, MPH was shown to enhance vulnerability to cocaine abuse and alter DAT kinetic parameters in PFC (Somkuwar et al. 2013). Adolescent MPH treatment increased adult cocaine self-administration in the SHR model of ADHD (Jordan et al. 2014). In contrast, MPH did not have effects on cocaine intake or cue reactivity in control rat strains (non-ADHD model rats).

There is significant human evidence to suggest that ADHD patients treated with stimulants do not have an increased risk of SUD above their already existing risk level (Upadhyaya 2007; Barkley et al. 2003; Wilson and Levin 2005). Based upon an examination of 443,041 patient records from the 2002–2009 NSDUH, NMU of prescription stimulants seemed to occur with individuals already engaged in broader patterns of drug abuse rather than NMU of stimulants triggering abuse of illicit drugs (Sweeney et al. 2013). Having ADHD is a high-risk factor for SUD, so adequately controlling ADHD symptoms with MPH may actually reduce the risk of developing SUD (Upadhyaya et al. 2010). There is support for this notion in that adult patients with greater ADHD symptom severity have been found to be more likely to misuse ADHD medications than those with lessor severity (Upadhyaya et al. 2010). In patients who already have comorbid ADHD and SUD, studies have generally found that ADHD medication, including MPH, provides no overall improvement in SUD symptoms (Upadhyaya 2007).

Overall, the idea that MPH medication use in ADHD patients does not lead to SUD is well-supported yet inconclusive and controversial, with some studies having important study limitations, including short periods of follow-up, high rates of attrition, and the inclusion of nicotine in the definition of SUD.

3.3 *Methylphenidate Treatment Benefit Versus Misuse Risk—Awareness Needed*

Despite the focus of this chapter on MPH misuse, it is important to note that the misuse issue must be addressed in the context of the overall benefit-risk assessment for using MPH in the treatment of ADHD. In most circumstances, the potential benefit likely outweighs the risk of misuse, with a plethora of evidence supporting the efficacy of MPH for the treatment of child (Katzman and Sternat 2014) and adult ADHD (Sopko et al. 2010). However, despite the overwhelming data suggesting MPH misuse is a health-care issue, a study of treatment of adult patients with ADHD by primary care physicians showed that >82 % of physicians did not suspect misuse of prescribed ADHD medication and <1 % believed that their patients were diverting prescribed ADHD medication (Lensing et al. 2013). These data suggest a lack of awareness or lack of acknowledgement of misuse issues. Thus, HCPs and patients need better education on stimulant misuse and its consequences, as well as improved monitoring for malingering and patient misuse to help avoid illicit diversion of these medications (Clemow and Walker 2014). When misuse or abuse are deemed a concern for a patient, monitoring of that patient's MPH use is important, and choosing an extended-release formulation of MPH, lisdexamfetamine, or a nonstimulant may be appropriate (Cortese et al. 2013;

Harstad et al. 2014). To reduce misuse, abuse, and diversion of MPH, it is important for HCPs to confirm an ADHD diagnosis, screen the patient for use of other drugs, provide guidance about appropriate and inappropriate medication use, and document the patient's prescription records (Harstad et al. 2014). It is particularly important to screen and monitor adolescents, because the risk for SUD is linked to having a history of NMU of prescription stimulants (McCabe and West 2013).

MPH diversion is not likely to go away as long as students seek drug-based ways to improve their perceived attention problems in hopes to better their academic performance (Rabiner et al. 2009b), and rates of ADHD prescriptions and malingering to get those prescriptions continue to rise (Jasinski et al. 2011; Sclar et al. 2012; Zuvekas and Vitiello 2012; Rabiner 2013). There is limited research, and none that is conclusive, on whether stimulants can enhance academic performance in those using it illicitly (Rabiner 2013), which is a point students should be educated about. Although not specifically studied, it is likely that students who delay studying because they expect the use of a stimulant will help them cram the night before an examination would actually perform better if they repetitively studied the material over time while under a less stressful setting (Clemow and Walker 2014). This theory fits with data showing that good study habits alone, even without stimulants, can overcome academic achievement disparity of ADHD patients compared with controls (Advokat et al. 2011). Students misusing MPH have been shown to be more likely to report lower grade point averages (McCabe et al. 2004b).

4 Conclusion

There are clinically relevant rates of MPH NMU, particularly in high school- and college-age students using the MPH to try to gain academic performance enhancement or using the MPH for recreational use. Such medication misuse can lead to deleterious side effects, particularly when taken with other recreational drugs such as alcohol, as evidenced by the number of annual ED visits linked to stimulant medication misuse. HCPs prescribing MPH should inquire about and educate themselves and their patients about MPH medication misuse, diversion, and the associated dangers. HCPs should also integrate strategies within their clinical practice to limit the potential for diversion and NMU of stimulant medications such as MPH. MPH is an important treatment option for patients suffering from ADHD, so if MPH is the HCPs treatment choice for their patient, consideration (based upon individual patient circumstances) should be given to long-acting rather than short-acting formulations, which show a significantly lower potential for misuse.

References

Advokat C, Lane SM, Luo C (2011) College students with and without ADHD: comparison of self-report of medication usage, study habits, and academic achievement. J Attn Disord 15 (8):656–666

Aldridge AP, Kroutil LA, Cowell AJ et al (2011) Medication costs to private insurers of diversion of medications for attention-deficit hyperactivity disorder. Pharmacoeconomics 29(7):621–635

American Psychiatric Association (2013) Diagnostic and statistical manual of mental disorders, 5th edn. American Psychiatric Publishing, Arlington

Andersen SL, Arvanitogiannis A, Pliakas AM et al (2002) Altered responsiveness to cocaine in rats exposed to methylphenidate during development. Nat Neurosci 5(1):13–14

Arias AJ, Gelernter J, Chan G et al (2008) Correlates of co-occurring ADHD in drug-dependent subjects: prevalence and features of substance dependence and psychiatric disorders. Addict Behav 33(9):1199–1207

Arnsten AFT (2011) Catecholamine influences on dorsolateral prefrontal cortical networks. Biol Psychiatry 69(12):e89–e99

Arnsten AFT, Li BM (2005) Neurobiology of executive functions: catecholamine influences on prefrontal cortical functions. Biol Psychiatry 57(11):1377–1384

Arnsten AFT, Rubia K (2012) Neurobiological circuits regulating attention, cognitive control, motivation, and emotion: disruptions in neurodevelopmental psychiatric disorders. J Am Acad Child Adolesc Psychiatry 51(4):356–367

Barkley RA, Fischer M, Smallish L et al (2003) Does the treatment of attention-deficit/hyperactivity disorder with stimulants contribute to drug use/abuse? A 13-year prospective study. Pediatrics 111(1):97–109

Barrett SP, Darredeau C, Bordy LE et al (2005) Characteristics of methylphenidate misuse in a university student sample. Can J Psychiatry 50(8):457–461

Berman SM, Kuczenski R, McCracken JT et al (2009) Potential adverse effects of amphetamine treatment on brain and behavior: a review (published correction appears in Mol Psychiatry. 2010;15(11):1121). Mol Psychiatry 14(2):123–142

Biederman J (2005) Attention-deficit/hyperactivity disorder: a selective overview. Biol Psychiatry 57(11):1215–1220

Biederman J, Faraone SV, Mick E et al (1995) High risk for attention deficit hyperactivity disorder among children of parents with childhood onset of the disorder: a pilot study. Am J Psychiatry 152(3):431–435

Biederman J, Monuteaux MC, Spencer T et al (2008) Stimulant therapy and risk for subsequent substance use disorders in male adults with ADHD: a naturalistic controlled 10-year follow-up study. Am J Psychiatry 165(5):597–603

Biederman J, Spencer TJ, Monuteaux MC et al (2010) A naturalistic 10-year prospective study of height and weight in children with attention-deficit hyperactivity disorder grown up: sex and treatment effects. J Pediatr 157:635–640

Biederman J, Wilens T, Mick E et al (1997) Is ADHD a risk factor for psychoactive substance use disorders? Findings from a four-year prospective follow-up study. J Am Acad Child Adolesc Psychiatry 36(1):21–29

Biederman J, Wilens T, Mick E et al (1999) Pharmacotherapy of attention-deficit/hyperactivity disorder reduces risk for substance use disorder. Pediatrics 104(2):e20

Bloom B, Jones LI, Freeman G (2013) Summary health statistics for U.S. children: national health interview survey, 2012. National center for health statistics. Vital Health Stat 10(258)

Bogle KE, Smith BH (2009) Illicit methylphenidate use: a review of prevalence, availability, pharmacology, and consequences. Curr Drug Abuse Rev 2(2):157–176

Booksh R, Pella RD, Singh AN et al (2010) Ability of college students to simulate ADHD on objective measures of attention. J Atten Disord 13(4):325–338

Brandon CL, Marinelli M, White FJ (2003) Adolescent exposure to methylphenidate alters the activity of rat midbrain dopamine neurons. Biol Psychiatry 54(12):1338–1344

Bredeloux P, Dubuc I, Costentin J (2007) Comparisons between bupropion and dexamphetamine in a range of in vivo tests exploring dopaminergic transmission. Br J Pharmacol 150(6):711–719
Bright GM (2008) Abuse of medications employed for the treatment of ADHD: results from a large-scale community survey. Medscape J Med 10(5):111
Bymaster FP, Katner JS, Nelson DL et al (2002) Atomoxetine increases extracellular levels of norepinephrine and dopamine in the prefrontal cortex of rat: a potential mechanism for efficacy in attention deficit/hyperactivity disorder. Neuropsychopharmacology 27(5):699–711
Carlezon WA Jr, Mague SD, Andersen SL (2003) Enduring behavioral effects of early exposure to methylphenidate in rats. Biol Psychiatry 54(12):1330–1337
Carlson GA, Kelly KL (2003) Stimulant rebound: how common is it and what does it mean? J Child Adolesc Psychopharmacol 13(2):137–142
Cascade E, Kalali AH, Weisler RH et al (2008) Seasonality and the changing adult/child prescription ratios in ADHD therapy. Psychiatry (Edgmont) 5(1):23–25
Cepeda MS, Fife D, Berwaerts J et al (2014) Shopping behavior for ADHD drugs: results of a cohort study in a pharmacy database. Drugs R D 14(3):205–211
Center for Medicare and Medicaid Services (2012) Drug diversion in the medicaid program: state strategies for reducing prescription drug diversion in medicaid. United States Department of Health and Human Services, Baltimore. http://www.cms.gov/Medicare-Medicaid-Coordination/Fraud-Prevention/MedicaidIntegrityProgram/downloads/drugdiversion.pdf. Accessed 26 Aug 2014
Charach A, Yeung E, Climans T et al (2011) Childhood attention-deficit/hyperactivity disorder and future substance use disorders: comparative meta-analyses. J Am Acad Child Adolesc Psychiatry 50(1):9–21
Chen LY, Strain EC, Alexandre PK et al (2014) Correlates of nonmedical use of stimulants and methamphetamine use in a national sample. Addict Behav 39(5):829–836
Christensen L, Sansané R, Hodgkins P et al (2010) Pharmacological treatment patterns among patients with attention-deicit/hyperactivity disorder: retrospective claims-based analysis of a managed care population. CMRO 26(4):977–989
Clemow DB, Walker DJ (2014) The potential for misuse and abuse of medications in ADHD: a review. Postgrad Med 126(5):64–81
Cohen AL, Jhung MA, Budnitz DS (2006) Stimulant medications and attention deficit-hyperactivity disorder. N Engl J Med 354(21):2294–2295
Concerta Package Insert (2010) McNeil Pediatrics, Titusville
Cooper WO, Habel LA, Sox CM et al (2011) ADHD drugs and serious cardiovascular events in children and young adults. N Engl J Med 365:1896–1904
Cortese S, Holtmann M, Banaschewski T et al (2013) European ADHD guidelines group. Practitioner review: current best practice in the management of adverse events during treatment with ADHD medications in children and adolescents. J Child Psychol Psychiatry 54(3):227–246
Dalsgaard S, Mortensen PB, Frydenberg M et al (2014) ADHD, stimulant treatment in childhood and subsequent substance abuse in adulthood—a naturalistic long-term follow-up study. Addict Behav 39(1):325–328
Darredeau C, Barrett SP, Jardin B et al (2007) Patterns and predictors of medication compliance, diversion, and misuse in adult prescribed methylphenidate users. Hum Psychopharmacol 22(8):529–536
Del Campo N, Chamberlain SR, Sahakian BJ et al (2011) The roles of dopamine and noradrenaline in the pathophysiology and treatment of attention-deficit/hyperactivity disorder. Biol Psychiatry 69(12):e145–157
Doremus-Fitzwater TL, Varlinskaya EI, Spear LP (2010) Motivational systems in adolescence: possible implications for age differences in substance abuse and other risk-taking behaviors. Brain Cogn 72(1):114–123
Dupont RL, Coleman JJ, Bucher RH et al (2008) Characteristics and motives of college students who engage in nonmedical use of methylphenidate. Am J Addict 17(3):167–171

Economidou D, Dalley JW, Everitt BJ (2011) Selective norepinephrine reuptake inhibition by atomoxetine prevents cue-induced heroin and cocaine seeking. Biol Psychiatry 69(3):266–274
Faraone SV, Biederman J, Morley CP et al (2008) Effect of stimulants on height and weight: a review of the literature. J Am Acad Child Adolesc Psychiatry 47:994–1009
Fayyad J, De Graaf R, Kessler R et al (2007) Cross-national prevalence and correlates of adult attention-deficit hyperactivity disorder. Br J Psychiatry 190:402–409
Fuermaier AB, Tucha L, Koerts J et al (2012) Measurement of stigmatization towards adults with attention deficit hyperactivity disorder 7(12):e51755
Gasior M, Bergman J, Kallman MJ et al (2005) Evaluation of the reinforcing effects of monoamine reuptake inhibitors under a concurrent schedule of food and i.v. drug delivery in rhesus monkeys. Neuropsychopharmacology 30(4):758–764
Gamo NJ, Wang M, Arnsten AFT (2010) Methylphenidate and atomoxetine enhance prefrontal function through α_2-adrenergic and dopamine D1 receptors. J Am Acad Child Adolesc Psychiatry 49(10):1011–1023
Groenman AP, Oosterlaan J, Rommelse NNJ et al (2013) Stimulant treatment for attention-deficit hyperactivity disorder and risk of developing substance use disorder. Br J Psychiatry 203:112–119
Habel LA, CooperWO Sox CM et al (2011) ADHD medications and risk of serious cardiovascular events in young and middle-aged adults. JAMA 306:2673–2683
Harp JP, Jasinski LJ, Shandera-Ochsner AL et al (2011) Detection of malingered ADHD using the MMPI-2-RF. Psychol Inj Law 4(1):32–43
Harrison AG, Edwards MJ, Parker KCH (2007) Identifying students faking ADHD: preliminary findings and strategies for detection. Arch Clin Neuropsychol 22(5):577–588
Harstad E, Levy S, and Committee on Substance Abuse (2014) Attention-deficit/hyperactivity disorder and substance abuse. Pediatrics 134(1):e293–e301
Heil SH, Holmes HW, Bickel WK et al (2002) Comparison of the subjective, physiological, and psychomotor effects of atomoxetine and methylphenidate in light drug users. Drug Alcohol Depend 67(2):149–156
Hutcheson DM, Everitt BJ (2003) The effects of selective orbitofrontal cortex lesions on the acquisition and performance of cue-controlled cocaine seeking in rats. Ann N Y Acad Sci 1003:410–411
Ito R, Dalley JW, Howes SR et al (2000) Dissociation in conditioned dopamine release in the nucleus accumbens core and shell in response to cocaine cues and during cocaine-seeking behavior in rats. J Neurosci 20(19):7489–7495
Ito R, Robbins TW, Everitt BJ (2004) Differential control over cocaine seeking behavior by nucleus accumbens core and shell. Nat Neurosci 7(4):389–397
Intuniv Package Insert (2011) Shire Inc, Wayne
Jachimowicz G, Geiselman RE (2004) Comparison of ease of falsification of attention deficit hyperactivity disorder diagnosis using standard behavioral rating scales. Cognitive Sci Online 2:6–20
Jasinski DR, Faries DE, Moore RJ et al (2008) Abuse liability assessment of atomoxetine in a drug-abusing population. Drug Alcohol Depend 95(1–2):140–146
Jasinski DR, Krishnan S (2009) Abuse liability and safety of oral lisdexamfetamine dimesylate in individuals with a history of stimulant abuse. J Psychopharmacol 23(4):419–427
Jasinski LJ, Harp JP, Berry DTR et al (2011) Using symptom validity tests to detect malingered ADHD in college students. Clin Neuropsychol 25(8):1415–1428
Joffe A (2014) Nonmedical use of prescription stimulants by adolescents. Adolesc Med 25 (1):89–103
Johnston LD, O'Malley PM, Bachman JG et al (2012) Monitoring the future: national survey results on drug use, 1975–2011: vol II, college students and adults ages 19–50. Institute for Social Research, The University of Michigan, Ann Arbor. http://www.monitoringthefuture.org/pubs/monographs/mtf-vol2_2011.pdf. Accessed 26 Aug 2014
Johnston LD, O'Malley PM, Bachman JG et al (2014) Monitoring the future: national survey results on drug use, 1975–2013: vol I, secondary school students. Institute for Social Research,

The University of Michigan, Ann Arbor. http://www.monitoringthefuture.org/pubs/monographs/mtf-vol1_2012.pdf. Accessed 26 Aug 2014
Jordan CJ, Harvey RC, Baskin BB et al (2014) Cocaine-seeking behavior in a genetic model of attention-deficit/hyperactivity disorder following adolescent methylphenidate or atomoxetine treatments. Drug Alcohol Depend 140:25–32
Kahlig KM, Binda F, Khoshbouei H et al (2005) Amphetamine induces dopamine efflux through a dopamine transporter channel. Proc Natl Acad Sci U S A 102(9):3495–3500
Kapvay Package Insert (2010) Shionogi Pharma Inc, Florham Park
Katusic SK, Barbaresi WJ, Colligan RC et al (2005) Psychostimulant treatment and risk for substance abuse among young adults with a history of attention-deficit/hyperactivity disorder: a population-based, birth cohort study. J Child Adolesc Psychopharmacol 15(5):764–776
Katzman MA, Sternat T (2014) A review of OROS methylphenidate (Concerta®) in the treatment of attention-deficit/hyperactivity disorder. CNS Drugs 28:1005–1033
Kawaura K, Karasawa J, Chaki S et al (2014) Stimulation of postsynapse adrenergic α2a receptor improves attention/cognition performance in an animal model of attention deficit hyperactivity disorder. Behav Brain Res 270:349–356
Kessler RC, Adler LA, Barkley R et al (2005) Patterns and predictors of ADHD persistence into adulthood: results from the national comorbidity survey replication. Biol Psychiatry 57 (11):1442–1451
Kessler RC, Adler L, Barkley R et al (2006) The prevalence and correlates of adult ADHD in the United States: results from the national comorbidity survey replication. Am J Psychiatry 163 (4):716–723
Klein-Schwartz W, McGrath J (2003) Poison centers' experience with methylphenidate abuse in pre-teens and adolescents. J Am Acad Child Adolesc Psychiatry 42(3):288–294
Koda K, Ago Y, Cong Y et al (2010) Effects of acute and chronic administration of atomoxetine and methylphenidate on extracellular levels of noradrenaline, dopamine and serotonin in the prefrontal cortex and striatum of mice. J Neurochem 114(1):259–270
Kollins SH (2003) Comparing the abuse potential of methylphenidate versus other stimulants: a review of available evidence and relevance to the ADHD patient. J Clin Psychiatry 64(Suppl 11):14–18
Kollins SH (2007) Abuse liability of medications used to treat attention-deficit/hyperactivity disorder (ADHD). Am J Addict 16(Suppl 1):35–42
Kollins SH, English J, Robinson R et al (2009) Reinforcing and subjective effects of methylphenidate in adults with and without attention deficit hyperactivity disorder (ADHD). Psychopharmacology 204(1):73–83
Kroutil LA, Van Brunt DL, Herman-Stahl MA et al (2006) Nonmedical use of prescription stimulants in the United States. Drug Alcohol Depend 84(2):135–143
Kuczenski R, Segal DS (1997) Effects of methylphenidate on extracellular dopamine, serotonin, and norepinephrine: comparison with amphetamine. J Neurochem 68(5):2032–2037
Lee SS, Humphreys KL, Flory K et al (2011) Prospective association of childhood attention-deficit/hyperactivity disorder (ADHD) and substance use and abuse/dependence: a meta-analytic review. Clin Psychol Rev 31(3):328–341
Lensing MB, Zainer P, Sandvik L et al (2013) Adults with ADHD: use and misuse of stimulant medication as reported by patients and their primary care physicians. Atten Defic Hyperact Disord 5(4):369–376
Lile JA, Stoops WW, Durell TM et al (2006) Discriminative-stimulus, self-reported, performance, and cardiovascular effects of atomoxetine in methylphenidate-trained humans. Exp Clin Psychopharmacol 14(2):136–147
Mao AR, Babcock T, Brams M (2011) ADHD in adults: current treatment trends with consideration of abuse potential of medications. J Psychiatr Pract 17(4):241–250
Mannuzza S, Klein RG, Truong NL et al (2008) Age of methylphenidate treatment initiation in children with ADHD and later substance abuse: prospective follow-up into adulthood. Am J Psychiatry 165(5):604–609

Marshall P, Schroeder R, O'Brien J et al (2010) Effectiveness of symptom validity measures in identifying cognitive and behavioral symptom exaggeration in adult attention deficit hyperactivity disorder. Clin Neuropsychol 24(7):1204–1237

McCabe SE, Teter CJ, Boyd CJ (2004a) The use, misuse and diversion of prescription stimulants among middle and high school students. Subst Use Misuse 39(7):1095–1116

McCabe SE, Teter CJ, Boyd CJ et al (2004b) Prevalence and correlates of illicit methylphenidate use among 8th, 10th, and 12th grade students in the United States, 2001. J Adolesc Health 35 (6):501–504

McCabe SE, Knight JR, Teter CJ et al (2005) Non-medical use of prescription stimulants among US college students: prevalence and correlates from a national survey (published corrections appears in Addiction. 2005; 100(4):573). Addiction 100(1):96–106

McCabe SE, Boyd CJ, Teter CJ (2006a) Medical use, illicit use, and diversion of abusable prescription drugs. J Am Coll Health 54(5):269–278

McCabe SE, Teter CJ, Boyd CJ (2006b) Medical use, illicit use and diversion of prescription stimulant medication. J Psychoactive Drugs 38(1):43–56

McCabe SE, West BT (2013) Medical and nonmedical use of prescription stimulants: results from a national multicohort study. J Am Acad Child Adolesc Psychiatry 52(12):1272–1280

McCabe SE, West BT, Teter CJ et al (2014) Trends in medical use, diversion, and nonmedical use of prescription medications among college students from 2003 to 2013: connecting the dots. Addict Behav 39:1176–1182

Merkel RL, Kuchibhatla A (2009) Safety of stimulant treatment in attention deficit hyperactivity disorder: part 1. Expert Opin Drug Saf 8(6):655–668

Messina BG, Silvestri MM, Diulio AR et al (2014) Alcohol use, impulsivity, and the non-medical use of prescription stimulants among college students. Addict Behavior 39:1798–1803

Molina BSG, Pelham WE Jr (2003) Childhood predictors of adolescent substance use in a longitudinal study of children with ADHD. J Abnorm Psychol 112(3):497–507

Murnane KS, Howell LL (2011) Neuroimaging and drug taking in primates. Psychopharmacology 216(2):153–171

Musso MW, Hill BD, barker AA et al (2014) Utility of the personality assessment inventory for detecting malingered ADHD in college students. J Attention Disord. doi:10.1177/1087054714548031 (Epub ahead of print 9 September 2014)

Musso MW, Gouvier WD (2014) "Why is this so hard?" A review of detection of malingered ADHD in college students. J Attention Discord 18(3):186–201

National Institute on Drug Abuse (2011) Commonly abused prescription drugs. http://www.drugabuse.gov/sites/default/files/rx_drugs_placemat_508c_10052011.pdf. Accessed 19 Aug 2014

Novak SP, Kroutil LA, Williams RL et al (2007) The nonmedical use of prescription ADHD medications: results from a national internet panel. Subst Abuse Treat Prev Policy 2:32

Olfson M, Huang C, Gerhard T et al (2012) Stimulants and cardiovascular events in youth with attention-deficit/hyperactivity disorder. J Am Acad Child Adolesc Psychiatry 51:147–156

Parasrampuria DA, Schoedel KA, Schuller R et al (2007) Do formulation differences alter abuse liability of methylphenidate? A placebo-controlled, randomized, double-blind, crossover study in recreational drug users. J Clin Psychopharmacol 27(5):459–467

Poulin C (2001) Medical and nonmedical stimulant use among adolescents: from sanctioned to unsanctioned use. CMAJ 165(8):1039–1044

Poulin C (2007) From attention-deficit/hyperactivity disorder to medical stimulant use to the diversion of prescribed stimulants to non-medical stimulant use: connecting the dots. Addiction 102(5):740–751

Quinn CA (2003) Detection of malingering in assessment of adult ADHD. Arch Clin Neuropsychol 18(4):379–395

Rabiner DL (2013) Stimulant prescription cautions: addressing misuse, diversion, and malingering. Curr Psychiatry Rep 15(7):375

Rabiner DL, Anastopoulos AD, Costello EJ et al (2009a) Motives and perceived consequences of nonmedical ADHD medication use by college students: are students treating themselves for attention problems? J Attent Disord 13(3):259–270
Rabiner DL, Anastopoulos AD, Costello EJ et al (2009b) The misuse and diversion of prescribed ADHD medications by college students. J Atten Disord 13(2):144–153
Rowlett JK, Platt DM, Yao WD et al (2007) Modulation of heroin and cocaine self-administration by dopamine D1- and D2-like receptor agonists in rhesus monkeys. J Pharmacol Exp Ther 321 (3):1135–1143
Rush CR, Essman WD, Simpson CA et al (2001) Reinforcing and subject-rated effects of methylphenidate and *d*-amphetamine in non-drug-abusing humans. J Clin Psychopharmacol 21 (3):273–286
Sansone RA, Sansone LA (2011) Faking attention deficit hyperactivity disorder. Innov Clin Neurosci 8(8):10–13
Sclar DA, Robison LM, Castillo LV et al (2012) Attention deficit/hyperactivity disorder among adults in the US: trend in diagnosis and use of pharmacotherapy. Pharmaceut Med 26(2):97–101
Sepúlveda DR, Thomas LM, McCabe SE et al (2011) Misuse of prescribed stimulant medication for ADHD and associated patterns of substance use: preliminary analysis among college students. J Pharm Pract 24(6):551–560
Schenk S, Izenwasser S (2002) Pretreatment with methylphenidate sensitizes rats to the reinforcing effects of cocaine. Pharmacol Biochem Behav 72(3):651–657
Schneider BN, Enenbach M (2014) Managing the risks of ADHD treatments. Curr Psychiatry Rep 16(10):479
Schramm-Sapyta NL, Cha YM, Chaudhry S (2007) Differential anxiogenic, aversive, and locomotor effects of THC in adolescent and adult rats. Psychopharmacology 191(4):867–877
Simpson D, Plosker GL (2004) Atomoxetine: a review of its use in adults with attention deficit hyperactivity disorder. Drugs 64(2):205–222
Sollman MJ, Ranseen JD, Berry DTR (2010) Detection of feigned ADHD in college students. Psychol Assess 22(2):325–335
Somkuwar SS, Jordan CJ, Kantak KM et al (2013) Adolescent atomoxetine treatment in a rodent model of ADHD: effects on cocaine self-administration and dopamine transporters in frontostriatal regions. Neuropsychopharmacology 38(13):2588–2597
Sopko MA Jr, Caberwal H, Chavez B (2010) The safety and efficacy of methylphenidate and dexmethylphenidate in adults with attention deficit/hyperactivity disorder. J Cent Nerv Syst Dis 2:15–30
Spencer TJ, Biederman J, Ciccone PE et al (2006) PET study examining pharmacokinetics, detection and likeability, and dopamine transporter receptor occupancy of short- and long-acting oral methylphenidate. Am J Psychiatry 163(3):387–395
Stahl SM (2003a) Neurotransmission of cognition, part 1, dopamine is a hitchhiker in frontal cortex: norepinephrine transporters regulate dopamine. J Clin Psychiatry 64(1):4–5
Stahl SM (2003b) Neurotransmission of cognition, part 2. Selective NRIs are smart drugs: exploiting regionally selective actions on both dopamine and norepinephrine to enhance cognition. J Clin Psychiatry 64(2):110–111
Stahl SM (2003c) Neurotransmission of cognition, part 3. Mechanism of action of selective NRIs: both dopamine and norepinephrine increase in prefrontal cortex. J Clin Psychiatry 64(3):230–231
Stahl SM (2008) Stahl's essential psychopharmacology: neuroscientific basis and practical application, 3rd edn. Cambridge University Press, New York
Stoops WW, Glaser PEA, Fillmore MT et al (2004) Reinforcing, subject-rated, performance and physiological effects of methylphenidate and *d*-amphetamine in stimulant abusing humans. J Psychopharmacol 18(4):534–543
Stoops WW, Lile JA, Fillmore MT et al (2005) Reinforcing effects of methylphenidate: influence of dose and behavioral demands following drug administration. Psychopharmacol 177:349–355

Strattera Package Insert (2014) Eli Lilly and Company, Indianapolis
Substance Abuse and Mental Health Services Administration (2001) Treatment for stimulant use disorders. Center for Substance Abuse Treatment, US Department of Health and Human Services, Rockville. https://store.samhsa.gov/shin/content/QGCT33/QGCT33.pdf. Accessed 26 Aug 2014
Substance Abuse and Mental Health Services Administration (2007) Results from the 2006 national survey on drug use and health: national findings. Office of Applied Studies, US Department of Health and Human Services, Rockville. https://www.asipp.org/documents/2006NSDUH.pdf. Accessed 26 Aug 2014
Substance Abuse and Mental Health Services Administration (2013) The DAWN report: emergency department visits involving attention deficit/hyperactivity disorder stimulant medications. Center for Behavioral Health Statistics and Quality, Rockville. http://www.samhsa.gov/data/2k13/dawn073/sr073-add-adhd-medications.htm. Accessed 26 Aug 2014
Suhr J, Hammers D, Dobbins-Buckland K et al (2008) The relationship of malingering test failure to self-reported symptoms and neuropsychological findings in adults referred for ADHD evaluation. Arch Clin Neuropsychol 23(5):521–530
Suhr JA, Sullivan BK, Rodriguez JL (2011) The relationship of noncredible performance to continuous performance test scores in adults referred for attention-deficit/hyperactivity disorder evaluation. Arch Clin Neuropsychol 26(1):1–7
Sullivan BK, May K, Galbally L (2007) Symptom exaggeration by college adults in attention-deficit hyperactivity disorder and learning disorder assessments. Appl Neuropsychol 14(3):189–207
Sweeney CT, Sembower MA, Ertischek MD et al (2013) Nonmedical use of prescription ADHD stimulants and preexisting patterns of drug abuse. J Addict Dis 32(1):1–10
Teter CJ, McCabe SE, Boyd CJ et al (2003) Illicit methylphenidate use in an undergraduate student sample: prevalence and risk factors. Pharmacotherapy 23(5):609–617
Teter CJ, McCabe SE, Cranford JA et al (2005) Prevalence and motives for illicit use of prescription stimulants in an undergraduate student sample. J Am Coll Health 53(6):253–262
Teter CJ, McCabe SE, LaGrange K et al (2006) Illicit use of specific prescription stimulants among college students: prevalence, motives, and routes of administration. Pharmacotherapy 26 (10):1501–1510
Tucha L, Sontag TA, Walitza S et al (2009) Detection of malingered attention deficit hyperactivity disorder. Atten Defic Hyperact Disord 1(1):47–53
Upadhyaya HP (2007) Managing attention-deficit/hyperactivity disorder in the presence of substance use disorder. J Clin Psychiatry 68(Suppl 11):23–30
Upadhyaya HP, Kroutil LA, Deas D et al (2010) Stimulant formulation and motivation for nonmedical use of prescription attention-deficit/hyperactivity disorder medications in a college-aged population. Am J Addict 19(6):569–577
Upadhyaya HP, Rose K, Wang W et al (2005) Attention-deficit/hyperactivity disorder, medication treatment, and substance use patterns among adolescents and young adults. J Child Adolesc Psychopharmacol 15(5):799–809
Volkow ND, Swanson JM (2003) Variables that affect the clinical use and abuse of methylphenidate in the treatment of ADHD. Am J Psychiatry 160(11):1909–1918
Volkow ND, Wang GJ, Fischman MW et al (1997) Relationship between subjective effects of cocaine and dopamine transporter occupancy. Nature 386(6627):827–830
Volkow ND, Wang GJ, Kollins SH et al (2009) Evaluating dopamine reward pathway in ADHD: clinical implications. JAMA 302(10):1084–1091
Volkow ND, Wang GJ, Newcorn J et al (2007a) Brain dopamine transporter levels in treatment and drug naïve adults with ADHD. Neuroimage 34(3):1182–1190
Volkow ND, Wang GJ, Newcorn J et al (2007b) Depressed dopamine activity in caudate and preliminary evidence of limbic involvement in adults with attention-deficit/hyperactivity disorder. Arch Gen Psychiatry 64(8):932–940
Wee S, Woolverton WL (2004) Evaluation of the reinforcing effects of atomoxetine in monkeys: comparison to methylphenidate and desipramine. Drug Alcohol Depend 75(3):271–276

Weiss M, Murray C (2003) Assessment and management of attention-deficit hyperactivity disorder in adults. CMAJ 168(6):715–722

Wetzell BB, Muller MM, Cobuzzi JL et al (2014) Effect of age on methylphenidate-induced conditioned taste avoidance and related BDNF/TrkB signaling in the insular cortex of the rat. Psychopharmacology 231(8):1493–1501

Whitelaw RB, Markou A, Robbins TW et al (1996) Excitotoxic lesions of the basolateral amygdala impair the acquisition of cocaine-seeking behaviour under a second-order schedule of reinforcement. Psychopharmacology 127(3):213–224

Wilens TE, Adamson J, Monuteaux MC et al (2008a) Impact of prior stimulant treatment for attention-deficit hyperactivity disorder in the subsequent risk for cigarette smoking, alcohol, and drug use disorders in adolescent girls. Arch Pediatr Adolesc Med 162(10):916–921

Wilens TE, Adler LA, Adams J et al (2008b) Misuse and diversion of stimulants prescribed for ADHD: a systematic review of the literature. J Am Acad Child Adolesc Psychiatry 47(1):21–31

Wilens TE, Faraone SV, Biederman J et al (2003) Does stimulant therapy of attention-deficit/hyperactivity disorder beget later substance abuse? A meta-analytic review of the literature. Pediatrics 111(1):179–185

Wilens TE, Gignac M, Swezey A et al (2006) Characteristics of adolescents and young adults with ADHD who divert or misuse their prescribed medications. J Am Acad Child Adolesc Psychiatry 45(4):408–414

Wilens TE, Martelon M, Joshi G et al (2011) Does ADHD predict substance use disorders? A 10-year follow-up study of young adults with ADHD. J Am Acad Child Adolesc Psychiatry 50 (6):543–553

Wilens TE, Morrison NR (2012) Substance-use disorders in adolescents and adults with ADHD: focus on treatment. Neuropsychiatry (London) 2(4):301–312

Wilson JJ, Levin FR (2005) Attention-deficit/hyperactivity disorder and early-onset substance use disorders. J Child Adolesc Psychopharmacol 15(5):751–763

Young CJ, Gross AM (2011) Detection of response bias and noncredible performance in adult attention-deficit/hyperactivity disorder. Arch Clin Neuropsychol 26(3):165–175

Zuvekas SH, Vitiello B (2012) Stimulant medication use among U.S. children: a 12-year perspective. Am J Psychiatry 169(2):160–166

Misuse and Associated Harms of Quetiapine and Other Atypical Antipsychotics

Mark E. Montebello and Jonathan Brett

Abstract Recent years have seen a significant increase in the reports of atypical antipsychotic diversion, misuse and even dependency syndrome. These reports have arisen amidst a marked increase in prescribing of these agents. Much of this increase in prescribing is because of a preferential use of these medications over typical antipsychotic agents to treat schizophrenia and bipolar disorder due to perceptions of fewer extrapyramidal side effects. However, there has also been a significant increase in the off-label prescribing of these medicines to treat less well evidence-based conditions. Misuse and abuse are perhaps surprising given the putative central role of dopamine in addiction and that these agents are dopamine antagonists. However, there may be other factors such as other pharmacological effects and increasing availability driving this misuse. It is also apparent that certain patient groups appear to be more at risk. Here, we explore the evidence behind the misuse of atypical antipsychotics with a focus of quetiapine. We consider the factors that may be driving this misuse, and then, we also detail some of the adverse effects that may ensue. We end by suggesting interventions at a prescriber and systems level that may be implemented to reduce the risk of atypical antipsychotic misuse.

Keywords Atypical antipsychotics · Quetiapine · Misuse · Dependence · Off-label prescribing

---

M.E. Montebello (✉)
Drug and Alcohol Service, South Eastern Sydney Local Health District, Surry Hills, Australia
e-mail: mark.montebello@sesiahs.health.nsw.gov.au

M.E. Montebello
National Drug and Alcohol Research Centre, University of New South Wales, Randwick, Australia

J. Brett
Drug Health Services, Sydney Local Health District, Camperdown, Australia

Curr Topics Behav Neurosci (2017) 34:125–139
DOI 10.1007/7854_2015_424
Published Online: 23 December 2017

Contents

1 Introduction

Since their introduction twenty years ago, atypical or second-generation antipsychotic medications such as quetiapine and olanzapine have superseded typical or first-generation antipsychotics in many countries including Australia. These medications have revolutionised the treatment of schizophrenia and bipolar disorder. They are also used for many off-label indications, some of which have a limited evidence base to support these practices. Off-label prescribing is the prescribing of a medication for an indication outside of that for which it is licensed. With such widespread use, there are growing concerns regarding harms related to antipsychotic use in certain populations. In particular, there is emerging evidence of diversion, misuse and dependence as well as morbidity and mortality related to overdose. This has come to the attention of researchers and the media (Milligan 2013) but is not well recognised by some prescribers and dispensers. Diversion is defined as the taking of prescription drugs, whether obtained by prescription or otherwise, other than in the manner or for the reasons or time period prescribed, or by a person for whom the drug was not prescribed (Crime UNOoDa 2011). Misuse is defined as the non-oral use of a prescribed medication or non-prescribed use of a medication (Reddel et al. 2014). The Tenth Revision of the International Classification of Diseases and Health Problems (ICD-10) defines the dependence syndrome as being a cluster of physiological, behavioural and cognitive phenomena in which the use of a substance or a class of substances takes on a much higher priority for a given individual than other behaviours that once had greater value. A central descriptive characteristic of the dependence syndrome is the desire (often strong, sometimes overpowering) to take the psychoactive drugs (which may or not have been medically prescribed), alcohol or tobacco. There may be evidence that return to substance use after a period of abstinence leads to a more rapid reappearance of other features of the syndrome than occurs with non-dependent individuals (WHO 2015).

2 What Is the Evidence of Misuse?

Quetiapine misuse is increasingly documented in case reports and case series, ambulance data and emergency department presentations with overdose, poison information centre data and coronial data.

2.1 Case Reports: Misuse, Abuse and Dependence

In 2008, Tcheremissine identified nine case reports and case series from January 1991 to July 2008 concerning the abuse and diversion of quetiapine in forensic populations and in individuals with histories of substance abuse. Tcheremissine (2008) concluded that the abuse potential of quetiapine must be carefully viewed in the context of its clinical relevance, the diversity of patient populations, the wide range of doses utilised and similar past concerns with other CNS-acting medications including other antipsychotics.

A review by Sansone and Sansone (2010) alerted clinicians in psychiatric and primary care settings to be vigilant to the potential risk, particularly in those observed to be at higher risk including males, in jail or inpatient psychiatric settings, and people with a past history of polysubstance abuse.

In 2014, Cubala et al. identified twenty-five case reports from 1966 to 2012 of quetiapine abuse/dependence in psychiatric patients including eight papers in the Tcheremissine systematic review. It was noted that a higher frequency of abuse and dependence was observed in men in their mid-thirties, and less than half of all patients had a history of substance abuse/dependence (Cubala and Springer 2014).

More recent case series have described quetiapine misuse in other populations including patients in a community-based methadone maintenance programme (McLarnon et al. 2012), an outpatient department of a tertiary care psychiatry hospital (Srivastava et al. 2013), an addiction treatment inpatient unit (Malekshahi et al. 2015) and from a survey of urban people who inject drugs (PWIDs) (Reddel et al. 2014).

In an addiction treatment inpatient unit, Malekshahi et al. screened 429 inpatients and found 73 (17.0 %) reported misuse of antipsychotics with alcohol, opioids, cocaine, methamphetamine and/or cannabis at any time and 39 (9.1 %) within the past year. Of the past year misusers, 25 (64.1 %) were interviewed, and their demographic profile matched many of the previous case studies, that is, most were male (76.0 %) polysubstance abusers (84.0 %). The most abused antipsychotic medication was quetiapine (96.0 %), and it was obtained primarily from doctors (52.0 %) and diversion from family/friends (48.0 %). When interviewed, the commonest reasons given for misusing antipsychotics included to "recover" from other substances (66.7 %), to "enhance" the effects of other substances (25.0 %) and to "experiment" (20.8 %). The most frequently reported positive effect was "feeling mellow" (75.0 %), and the most frequently reported negative effects were consistent with antipsychotic use, e.g. feeling thirsty and trouble concentrating. Because of this prevalent misuse, Malekshahi et al. (2015) recommended that physicians should assess for misuse of atypical antipsychotics among patients with addiction issues.

In the self-report survey of 868 Australian urban PWID, 31 % of the sample [95 % confidence interval (CI) = 28–34 %] reported ever misusing quetiapine and 15 % (95 % CI = 13–17 %) in the preceding 6 months. Recent misuse was associated with localities reporting a higher rate of prescriptions. Recent quetiapine misuse was also associated with violent crime in the preceding month [odds ratio (OR) = 1.96, 95 % CI = 1.17–3.29]; non-heroin drug overdose in the preceding 12 months (OR = 3.52, 95 % CI = 1.39–8.91); non-prescribed benzodiazepine use (OR = 4.26, 95 % CI = 2.06–8.82); non-prescribed pharmaceutical opioid use (OR = 2.76, 95 % CI = 1.47–5.19); and amphetamine use (OR = 2.08, 95 % CI = 1.02–4.22) in the previous 6 months (Reddel et al. 2014).

There are a limited number of case reports for misuse for other atypical antipsychotics, with most of them involving olanzapine (Kumsar and Erol 2013; Lai 2010; Reeves 2007). Many of these reports are similar to those of quetiapine misuse. Lai hypothesised that olanzapine is at risk of abuse because of its modulating role in the dopaminergic reward system, unlike aripiprazole because it is a D2 and 5HT 1a partial agonist, with higher D2/D3 receptor binding in extrastriatal compared with striatal receptors, with consequent dopamine stabilising effects (Lai 2010).

In summary, there is increasing evidence from population-based surveys and case reports/series of atypical antipsychotic misuse, particularly with quetiapine. Higher-risk patients are male, have a history of substance misuse and are seen in forensic, psychiatric or addiction treatment settings. Atypical antipsychotics may be prescribed, and there is emerging evidence of diversion. The route of administration includes swallowing, inhaling and injecting. Almost all report psychological dependence, most report discontinuation effects, and some report reinforcing properties, and hence, it may be difficult to differentiate these from physiologically dependent withdrawal. The potential psychoactive effects of psychiatric medications may have been under-recognised, and further research is required to delineate the acute and long-term effects of these reports (Moncrieff et al. 2013). Most authors recommend caution when prescribing quetiapine to patients with a history of substance abuse/dependence and close monitoring for medication misuse during the course of treatment.

2.2 *Ambulance Data and Emergency Department Presentations with Overdose*

A study in Victoria, Australia, compared quetiapine-, olanzapine- and risperidone-related ambulance attendances in metropolitan Melbourne and prescription data in Victoria, Australia. This showed that quetiapine was consistently associated with higher rates of ambulance attendances relative to prescription availability than olanzapine or risperidone. Quetiapine-related attendances were associated with concurrent heroin and opioid replacement therapy toxicity, history of heroin and alcohol misuse, mood disorders, low Glasgow Coma Scale and women (Heilbronn et al. 2013).

The Hunter Area Toxicology Service database is one of the largest poisoning cohorts in the world. A recent study showed that between 1987 and 2012, overdoses with atypical antipsychotics increased more than any other substances (from 0 to 9.5 % of overall overdoses) (Buckley et al. 2015).

2.3 *Poison Information Centre Data*

A recent analysis of American Association of Poison Control Centers National Poison Data System data from 2005 to 2011 on single substance quetiapine exposures showed that misuse and abuse were prevalent in the general population. Intentional abuse was defined as improper or incorrect use of a substance in an attempt to gain a high, euphoric effect or some other psychotropic effect and included recreational use of a substance. Intentional misuse was defined as the improper or incorrect use of a substance for reasons other than psychotropic effects, such as deliberately increasing the dosage to enhance its effect. There were 3116 cases meeting inclusion criteria for quetiapine misuse (1948 cases) and abuse (1168 cases). The median age was 23 years. Misuse was reported most often in adults, whereas abuse occurred most frequently in adolescents. There were no deaths, but moderate or major toxicity occurred in 23.7 and 27.1 % of misuse and abuse cases, respectively. Seventy-six per cent were treated in the emergency department and/or received medical admission (Klein-Schwartz et al. 2014).

2.4 *Coronial Data*

A study of coronial data in Victoria, Australia, between 2001 and 2009 identified 224 deaths associated with quetiapine, with an average age of 43 years (range 15–87 years) (Pilgrim and Drummer 2013). The cause of death was mostly drug toxicity ($n = 114$, 51 %), followed by natural disease ($n = 60$, 27 %), external injury

(n = 31, 14 %) and unascertained causes (n = 19, 8 %). Depression and/or anxiety were common, observed in over a third of the cohort (n = 80, 36 %). About 20 % of cases did not mention a psychiatric diagnosis at all, which raises the possibility of off-label use of quetiapine. There were only a small number of cases of deaths in which quetiapine was not prescribed, possibly indicating the lack of an association between quetiapine diversion and death. However, fatalities in drug users associated with atypical antipsychotics have been reported (Simonsen et al. 2015).

2.5 *Other Potential Sources of Information*

There are a number of other methods for detecting misuse and abuse of substances that have predominantly been applied to prescription opioids. In Australia, the 2013 National Household Drug Survey highlighted significant rates of misuse of pharmaceuticals, but atypical antipsychotics are not included in these surveys (Welfare AIoHa 2013). Also from Australia in 2013, 45 % of the Illicit Drug Reporting System sample reported ever using quetiapine, with 18 % using in the past six months. This was a significant decrease from 25 % in 2012 (Stafford and Burns 2014). Other systems such as police seizure data, sentinel group surveys of drug users and college students, crowd-sourced information sites (such as www.streetRx.com), Pharmaceutical Benefits Scheme (PBS) prescription analysis and Web monitoring could all be used to monitor for antipsychotic abuse.

3 Case example 1

Illustrating off-label prescribing and subsequent significant morbidity.

Miss T is a 28-year-old woman who lives with a flatmate and is studying marketing at university. She had a father who was alcohol dependent and a mother who was the victim of family violence. Although she has no formal mental health diagnoses, she finds her emotions can be overwhelming. She has always had trouble sleeping and has been under a lot of pressure at university. She did take zopiclone in the past (prescribed by a GP across town) to help her sleep but ended up having to take increasing amounts to get to sleep and admits occasionally taking them through the day to help with anxiety and so she is worried about taking these again. She insists she needs something to sleep or she is going to go crazy. She is prescribed quetiapine 100 mg at night and given one month's supply. The following night she breaks up with her boyfriend and cannot sleep, so after drinking half a bottle of vodka, she takes all of the quetiapine tablets. Miss T is found by her flatmate 4 h later unconscious on the sofa with vomit in her mouth. She requires hospitalisation including a prolonged admission to intensive care to treat aspiration pneumonia.

4 Does the Pharmacological Properties of Quetiapine also Increase Misuse?

In general, substances of misuse and dependence cause activation of the dopaminergic mesolimbic reward pathway. Since atypical antipsychotics are dopaminergic receptor antagonists, it is counterintuitive that they may be substances of misuse and dependence unless other receptor systems are involved. In general, atypical antipsychotics have greater affinity for 5HT 2a compared to D2 receptors. Quetiapine is no exception but also acts on other receptors including H1 receptors, M1 acetylcholine 5HT 1a, 5HT 2c and alpha-adrenergic. At low doses (25 mg), quetiapine acts mostly as a H1 receptor antagonist; at moderate doses (50–200 mg), quetiapine increasingly antagonises serotonin receptors; at high doses (300 mg or more), quetiapine increasingly antagonises dopamine receptors (Seroquel 2013). There are several postulated mechanisms of action for the rationale for quetiapine misuse including sedation (H1) (Sansone and Sansone 2010; Smith 1980; Kuroki et al. 2008) and anxiolytic effects (alpha-adrenergic) (Sansone and Sansone 2010). Quetiapine also rapidly dissociates from the D2 receptor, possibly through relatively lower potency and decreased residence time; this also reduces the risk of extrapyramidal side effects (Kuroki et al. 2008; Bogart and Ott 2011).

5 What Other Factors Are Driving Misuse?

5.1 Prescribing Practices

Over the past century, many parts of the world have experienced successive waves of pharmaceutical medication misuse. These have included “over-the-counter” heroin available until the 1910s, unregulated amphetamines until the 1960s and barbiturates until the 1970s. Benzodiazepines superseded barbiturates in the 1960s because of their relative safety in overdose and the belief they were “non-addictive”. By the 1980s, the potential for benzodiazepines to be diverted, misused or cause dependence was well recognised. In Australia in the past decade, benzodiazepine prescription rates have only slightly declined, while opioids and atypical antipsychotics prescription rates have both significantly increased [see Fig. 1—overall dispensing for the six main classes of psychotropic drugs from 2000 to 2011 (Stephenson et al. 2012)].

Possible reasons for the increase in prescribing of atypical antipsychotics include increased acceptance of diagnosis and treatment of psychiatric disorders, difficulties in accessing psychosocial services and hence the reliance on psychotropic medication and a range of off-label uses for these medications. The potential for prescription opioids to be diverted, misused or cause dependence is well recognised, and now there is increasing evidence that atypical antipsychotics can cause similar problems.

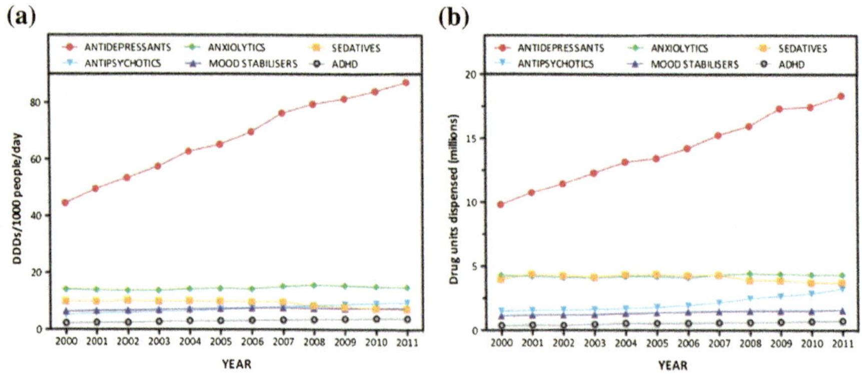

Fig. 1 Overall dispensing for the six main classes of psychotropic drugs from 2000 to 2011, represented in terms of **a** defined daily doses per 1000 population per day and **b** number of prescriptions dispensed

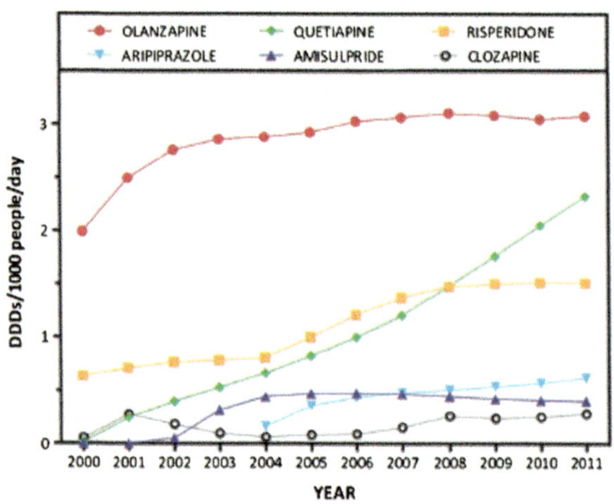

Fig. 2 Atypical antipsychotics: dispensing data for the six leading atypical antipsychotics from 2000 to 2011

Over the past fifteen years, the atypical antipsychotics with the greatest market share in Australia are olanzapine, quetiapine and risperidone [see Fig. 2—number of dispensing episodes for antipsychotics (Stephenson et al. 2012)].

5.2 *Off-Label Prescribing*

In Australia, licensed indications for medicines are designed to regulate the claims that can be made about a medicine by a pharmaceutical company. The Australian PBS lists the indications for quetiapine as follows:

- Schizophrenia (authority required—streamlined 1589);
- Monotherapy, for up to 6 months, of an episode of acute mania associated with bipolar I disorder (authority required—streamlined 2765);
- Maintenance treatment of bipolar I disorder (authority required—streamlined 2044).

In many parts of the world, off-label prescribing is legal and common and gives clinicians scope to treat patients who are refractory to standard therapy or where there is no licensed medication for an indication (McKean and Monasterio 2012). Quetiapine is prescribed off-label for many disorders including the following:

- Augmentation in major depressive disorder;
- Obsessive–compulsive disorder;
- Post-traumatic stress disorder;
- Insomnia;
- Behavioural disturbance in dementia;
- Borderline personality disorder;
- Alcohol dependence;
- Cannabis withdrawal.

In Australia, there are concerns regarding the high number of prescriptions for quetiapine 25 mg strength tablets. This is not a therapeutic dose to treat the PBS-subsidised indications of schizophrenia and bipolar disorder. In adults aged 20–59 years, this may indicate use for off-label indications such as for insomnia. In those over 75 years of age, this medication is also being used to treat behavioural disturbances in dementia. These off-label uses expose recipients to additional harm with little benefit and come at a significant financial cost to the PBS (Hollingworth et al. 2010). Consequently, a number of initiatives have been implemented to support better medication use including reducing the inappropriate use of atypical antipsychotic medications for sedation (Service NP 2014).

6 Potential Risks Versus Benefits of Increased Prescribing of Atypical Antipsychotics

The increased use of atypical antipsychotics exposes a larger population to their potential adverse effects, despite variable levels of evidence for their efficacy or safety for some conditions, and often does not represent quality use of medications. The risks and benefits of off-label prescribing should be carefully considered, and an informed decision is made by the patient before prescribing. The potential risks and benefits to the individual may be difficult to determine because of a lack of good evidence, particularly in people with comorbid substance-use and mental health disorders including schizophrenia and bipolar disorder. These patients are often excluded from original registration or other clinical trials despite high rates of comorbidity. For instance, more than half of all individuals with bipolar disorder have a substance-use disorder at some point in their lifetime (Kessler et al. 2005). Off-label prescribing may mean that people vulnerable to drug misuse are prescribed antipsychotics and therefore put at risk. Increase in off-label prescribing also increases the availability of antipsychotics in the community at large, therefore increasing the potential for diversion and misuse. The reasons why clinicians continue to prescribe atypical antipsychotics for poorly evidence-based indications are unclear but may include a lack of feasible alternatives, a lack of understanding of the evidence and risk, regulatory, industry and professional incentives as well as a host of other physician- and patient-related factors (Scott and Elshaug 2013).

7 What Are the Harms Associated with Misuse?

Misuse of atypical antipsychotics is associated with the known adverse effects of quetiapine, as well as overdose and cost to health systems.

Side effects Side effects are extrapyramidal including akathisia, orthostatic hypotension, somnolence, increased appetite, weight gain, tachycardia, palpitations, blurred vision, vomiting, peripheral oedema, abnormal dreams, dry mouth, dizziness and constipation.

QT prolongation The risk appears to be unique to certain atypical antipsychotic agents. Amisulpride is best known for this effect and in overdose can increase the risk of torsades de point (TdP), although there is little evidence that this agent is subject to misuse. Quetiapine is also associated with QT prolongation, and patients may be at increased risk of TdP if they have concomitant illness including cardiovascular disease, family history of long QT syndrome(or sudden cardiac death) and/or taking medications known to cause electrolyte imbalance or increase QT interval, including class Ia antiarrhythmics (e.g. disopyramide) or class III antiarrhythmics (e.g. amiodarone, sotalol), antipsychotic medications (e.g. amisulpride, ziprasidone, chlorpromazine, haloperidol), antibiotics (e.g. moxifloxacin, erythromycin) or any other class of medications known to prolong the QT interval (e.g. citalopram, pentamidine, methadone).

Long-term effects Longerm effects include extrapyramidal symptoms such as tardive dyskinesia, weight gain, hyperglycaemia and diabetes mellitus, increases in triglycerides and cholesterol, and decreases in fasting HDL cholesterol.

Uncommon high-risk events These include sudden cardiac death (particularly in the elderly), neuroleptic malignant syndrome, neutropenia and agranulocytosis (AstraZeneca 2014).

Discontinuation syndrome Discontinuation syndrome includes nausea, insomnia, headache, diarrhoea, vomiting, dizziness and irritability. There is debate on how this is different to withdrawal syndromes associated with other substance-use disorders (Tcheremissine 2008; Nielsen et al. 2012).

Table 1 Financial cost of antipsychotics in Australia in 2015

Medication	Patent	Number of tablets	Cost AUD$
Atypical antipsychotics			
Quetiapine 25 mg	Off patent 2012	60	$29.23
Quetiapine 200 mg	Off patent 2012	60	$89.63
Quetiapine XR 50 mg	On patent	60	$50.33
Quetiapine XR 200 mg	On patent	60	$89.63
Aripiprazole 20 mg	On patent	30	$253.77
Olanzapine 10 mg tablets	Off patent	28	$93.91
Risperidone 2 mg tablets	Off patent	60	$45.70
Typical antipsychotics			
Chlorpromazine 100 mg	Off patent	100	$17.78
Haloperidol 5 mg	Off patent	50	$11.02

Overdose Quetiapine-related overdoses were more likely to result in hypotension, respiratory depression, coma or death than all the other antipsychotics combined (Balit et al. 2003). The risk of intubation is associated with the dose ingested, and the evidence suggests that there is no increased risk of TdP with acute single-agent overdose (Isbister and Duffull 2009).

Financial cost Compared to other classes of medications including typical antipsychotics, atypical antipsychotics are expensive medications—see Table 1 (Health AGDo2015).

If a patient has a PBS concession card, they would pay $6.10 for an authority script of sixty quetiapine 100 mg tablets. In Sydney in 2015, the street value of one quetiapine 100 mg tablet is up to $5.00, hence increasing the incentive for diversion to the black market.

Studies in other countries have also demonstrated the financial costs of atypical antipsychotics to health systems. In New Zealand in 2010, the off-label use of quetiapine alone accounted for an estimated 17 % of budget for atypical antipsychotics (McKean and Monasterio 2012). An audit of the prescribing patterns of New Zealand general psychiatrists found that 60 % of inpatients and 62 % of community patients were prescribed medications to aid with sleep. Less zopiclone and benzodiazepines were prescribed than in other studies, and 60 % of the community prescriptions for quetiapine were primarily for sedation (Huthwaite et al. 2014).

8 Case example 2

This illustrates pressures in prescribing in an Australian general practice setting.

Mr J is a 38-year-old single long-distance truck driver who presents to a general practitioner requesting quetiapine for schizophrenia. He tells the doctor that he is not psychotic and that he needs a script because his usual general practitioner has moved interstate. Mr J is vague about past psychiatric symptoms and denies past hospitalisations. He does not disclose that he smokes 0.3 g of methamphetamine on most weekends and uses quetiapine and/or benzodiazepines for the comedown period. The doctor is suspicious about Mr J's history but feels pressured to prescribe because of the risk of relapse. The doctor contacts the Pharmaceutical Benefits Services Hotline but is declined an authority script because Mr J has been issued two recent scripts for quetiapine. Mr J is angry when he leaves the doctor because he is unable to obtain another script for quetiapine.

9 Responses to Reduce Harms from Atypical Antipsychotics

Responses to the misuse of these prescription medicines can be adapted from past responses to other prescription medications including benzodiazepines and opioids (Australia Co2012).

9.1 At the prescriber level

1. Assessment and treatment of comorbid disorders using evidence-based treatment and avoiding prescribing for conditions that do not have a good evidence base instead of using a more evidence-based approach, e.g. cognitive behavioural therapy; relaxation techniques; exercise; diet; and psychosocial supports.
2. Use of universal precautions when prescribing including a risk assessment of the potential for misuse, defined goals of therapy, a dose ceiling and plan for medication review.
3. Open dialogue with patients about the potential risks and benefits of medication use and risks of misuse.
4. Incorporation of addiction medicine principles into mainstream health care such as monitoring for aberrant drug behaviours using the yellow and red flags model (see Table 2).

9.2 At a systems level

1. Improved treatment models for chronic diseases and conditions. Trying to solve complex chronic medical, psychological and social conditions with medications alone is unrealistic and does not work. Patient-centred, multidisciplinary approach with coordinated care between providers is the best treatment model.

Table 2 Monitoring for aberrant drug behaviours (Passik et al. 1998a; Passik et al. 1998b)

Yellow flags	Red flags
Few unsanctioned dose escalations	Many unsanctioned dose escalations
Occasionally getting from multiple doctors	Often acquiring from multiple doctors
Aggressive complaining about need for higher dose or specific drug	Recurrent prescription loss
Drug hoarding during periods of reduced symptoms	Obtaining prescription drugs from non-medical sources
Unapproved use of drug to treat other symptoms (e.g. sleep, mood)	Concurrent substance abuse (alcohol/other drugs)
Reporting psychic effects not intended by clinician	Injecting oral formulations
	Selling prescription drugs
	Prescription forgery

Understanding of the risks and the benefits of off-label prescribing for given indications and being able to discuss this information with a patient is an important element of the quality use of medicines. This is often challenging given the paucity of evidence (rather than evidence of ineffectiveness) for some of these practices, and there is a growing need to fill this evidence gap. In the meantime, improving quality of prescribing would putatively reduce availability in the community and potentially reduce diversion. Local and international guidelines including the Australian Therapeutic Guidelines and the Choosing Wisely programme (implemented by the National Prescribing Service in Australia) can be used to guide this prescribing. For example, the implementation of guidelines and education to prescribers treating insomnia in inmates in a US prison reduced prescribing of low-dose quetiapine by 59 % after 22 months (Reeves 2012).

2. Limits to promotion and advertising by the pharmaceutical industry.
3. Identifying and responding to high-risk groups.
4. Sharing of electronic medical records between healthcare providers.
5. Prescription of monitoring systems.

10 Conclusions

Typical antipsychotics have been superseded by atypical antipsychotic medications including quetiapine and olanzapine. There is emerging evidence of a range of harms with these medications from case reports/series, ambulance data, emergency department presentation data, poison information centre data and coronial data. These harms include diversion, misuse and dependence as well as morbidity and mortality related to overdose. Populations at a higher risk of these harms include males, those with a history of substance misuse and those seen in forensic, psychiatric or addiction treatment settings. Almost all cases report psychological dependence, most report discontinuation effects, and some report reinforcing properties of these misused medications. Given the dopamine hypothesis of addiction, it is unusual that dopamine antagonists may be subject to misuse and dependence syndrome. However, there may be other receptor systems that these agents interact with to produce these effects. The burgeoning off-label prescribing of these atypical antipsychotic agents is increasing their availability in the public, thus increasing the risk of diversion and misuse. Responses to reduce harms from atypical antipsychotics include a thorough risk assessment for substance-use disorder before initiating and close monitoring for medication misuse during the course of treatment and improved treatment models for chronic diseases.

References

Australia Co (2013) National pharmaceutical drug misuse framework for action (2012–2015)

Balit CR, Isbister GK, Hackett LP, Whyte IM (2003) Quetiapine poisoning: a case series. Ann Emerg Med 42:751–758

Bogart GT, Ott CA (2011) Abuse of second-generation antipsychotics: what prescribers need to know. Curr Psychiatry 10:77–79

Buckley N, Whyte IM, Dawson AH, Isbister GK (2015) 26 Years of self-poisoning in Newcastle, 1987–2012. Aust Med J 202(8):438–442

Crime UNOoDa (2011) The non-medical use of prescription drugs. In: Policy direction issues discussion paper. United Nations Office on Drugs and Crime, New York

Cubala WJ, Springer J (2014) Quetiapine abuse and dependence in psychiatric patients: a treatment. J Subst Use 19:388–393

Health AGDo (2015) Schedule of pharmaceutical benefits. Australian Capital Territory: Commonwealth of Australia

Heilbronn CE, Lloyd B, McElwee P, Eade A, Lubman DI (2013) Trends in quetapine use and non-fatal quetiapine-related ambulance attendances. Drug Alcohol Rev 32:405–411

Hollingworth SA, Siskind DJ, Nissen LM, Robinson M, Hall WD (2010) Patterns of antipsychotic medication use in Australia 2002–2007. Aust NZJ Psychiatry 44:372–377

Huthwaite M, Cleghorn M, MacDonald J (2014) 'Out of the frying pan': the challenges of prescribing for insomnia in psychiatric patients. Australas Psychiatry 22:288–291

Isbister GK, Duffull SB (2009) Quetiapine overdose: predicting intubation, duration of ventilation, cardiac monitoring and the effect of activated charcoal. Int Clin Psychopharmacol 24:174–180

Kessler RC, Chiu WT, Demler O, Walters EE (2005) Prevalence, severity and comorbidity of 12-Month DSM-IV disorders in the national comorbidity survey replication. Arch Gen Psychiatry 62:617–627

Klein-Schwartz W, Schwartz EK, Anderson BD (2014) Evaluation of quetiapine abuse and misuse reported to poison centers. J Addict Med 8:195–198

Kumsar NA, Erol A (2013) Olanzapine abuse. Subst Abuse 34:73–74

Kuroki T, Nagao N, Nakahara T (2008) Neuropharmacology of second-generation antipsychotic drugs: a validity of the serotonin-dopamine hypothesis. Prog Brain Res 172:199–212

Lai CH (2010) Olanzapine abuse was relieved after switching to aripiprazole in a patient with psychotic depression. Prog Neuropsychopharmacol Biol Psychiatry 34:1363–1364

Malekshahi T, Tioleco N, Ahmed N, Campbell ANC, Haller D (2015) Misuse of atypical antipsychotics in conjunction with alcohol and other drugs of abuse. J Subst Abuse Treat 48:8–12

McKean A, Monasterio E (2012) Off-label use of atypical antipsychotics: cause for concern? CNS Drugs 26:383–390
McLarnon ME, Fulton HG, MacIsaac C, Barrett SP (2012) Characteristics of quetiapine misuse among clients of a community-based methadone maintenance program. J Clin Psychopharmacol 32:721–723
Milligan L (2013) Concerns grow over top-selling drug's side effects. In: Corporation AB (ed) 730 Report: ABC
Moncrieff J, Cohen D, Porter S (2013) The psychoactive effects of psychiatric medication: the elephant in the room. J Psychoactive Drugs 45:409–415
Nielsen M, Hansen EH, Gøtzsche PC (2012) What is the difference between dependence and withdrawal reactions? A comparison of benzodiazepines and selective serotonin re-uptake inhibitors. Addiction 107:900–908
Passik SD, Portenoy RK, Ricketts PL (1998a) Substance abuse issues in cancer patients. Part 1: prevalence and diagnosis. Oncology 12:517–521
Passik SD, Portenoy RK, Ricketts PL (1998b) Substance abuse issues in cancer patients. Part 2: evaluation and treatment. Oncology 12:729–734
PBS Schedule search for "QUETIAPINE" (2015) Commonwealth of Australia. Accessed 12/02/2015, at http://www.pbs.gov.au/pbs/search?term=QUETIAPINE
Pilgrim JL, Drummer OH (2013) The toxicity and comorbidities of fatal cases involving quetiapine. Forensic Sci Med Pathol 9:170–176
Reddel SE, Bruno R, Burns L, Kirwan A, Lukuge K, Dietze P (2014) Prevalence and associations of quetiapine fumarate misuse among an Australian national city sample of people who regularly inject drugs. Addiction 109:295–302
Reeves RR (2007) Abuse of Olanzapine by substance users. J Psychoactive Drugs 39:297–299
Reeves R (2012) Guideline, education and peer comparison to reduce prescriptions of benzodiazepines and low-dose quetiapine in prison. J Corrective Health Care 18:45–52
Sansone RA, Sansone LA (2010) Is seroquel developing an illicit reputation for misuse/abuse? Psychiatry 7:13–16
Scott IA, Elshaug AG (2013) Foregoing low-value care: how much evidence is needed to change beliefs? Intern Med J 43:107–109
Seroquel (quetiapine fumarate) Tablets (2013) AstraZeneca Pharamceuticals LP. Accessed 14/05/2015, at http://www1.astrazeneca-us.com/pi/seroquel.pdf
Seroquel XR quetiapine fumarate Product Information. AstraZeneca, 2014. Accessed 14/05/2015, at http://www.medicines.org.au/files/appserxr.pdf
Service NP (2014) Low-dose quetiapine: place in therapy? Health News and Evidence: NPS Medicinewise
Simonsen KW, Edvardsen HME et al (2015) Fatal poisoning in drug addicts in the Nordic countries in 2012. Forensic Sci Int 248:172–180
Smith JM (1980) Abuse of the antiparkinson drugs: a review of the literature. J Clin Psychiatry 41:351–354
Srivastava A, Patil V, Da Silva Pereira Y (2013) A case series of quetiapine addiction/dependence. Ger J Psychiatry 16:152–155
Stafford J, Burns L (2014) Australian Drug Trends 2013. In: Findings from the illicit drug reporting system (IDRS). University of New South Wales, Australia
Stephenson CP, Karanges E, McGregor IS (2012) Trends in the utilisation of psychotropic medications in Australia from 2000 to 2011. Aust NZJ Psychiatry 47:74–87
Tcheremissine OV (2008) Is quetiapine a drug of abuse? Re-examining the issueof addiction. Expert Opin Drug Safety 7:739–748
Welfare AIoHa (2013) National drug strategy household survey detailed report. Canberra
WHO (2015) Management of substance abuse. Accessed 28/05/2015, at http://www.who.int/substance_abuse/terminology/definition1/en/

Benzodiazepines

Suzanne Nielsen

Abstract Benzodiazepines have been in clinical use since the 1960s. Benzodiazepines act through allosteric modulation of the $GABA_A$ receptor to enhance the activity of GABA, an inhibitory neurotransmitter, resulting in a slowing of neurotransmission and sedative and anxiolytic effects. Initially benzodiazepines were thought to have low dependence liability, though over time there has been increasing evidence of benzodiazepine dependence. Benzodiazepines are commonly used to treat anxiety and insomnia, though increasingly they are considered second line treatments for most indications. Concerns about the effects of benzodiazepines on cognition, falls and their implication in opioid related mortality have emerged. Few pharmacological treatments for benzodiazepine dependence have been shown to be effective with gradual taper the most common treatment strategy for benzodiazepine dependence.

Keyword Benzodiazepine · Misuse · Non-medical use · Dependence

Contents

S. Nielsen (✉)
National Drug and Alcohol Research Centre, UNSW, 22–32 King Street, Randwick, NSW 2031, Australia
e-mail: suzanne.nielsen@unsw.edu.au

S. Nielsen
South East Sydney Local Health District (SESLHD) Drug and Alcohol Services, 591-623 S Dowling St, Surry Hills, NSW 2010, Australia

Curr Topics Behav Neurosci (2017) 34:141–159
DOI 10.1007/7854_2015_425
Published Online: 23 December 2017

1 History of Benzodiazepines

Benzodiazepines have been in clinical use as sedatives/hypnotic drugs since the 1960s, with early benzodiazepines such as chlordiazepoxide and diazepam being used for their improved safety profile over barbiturates (Wick 2013; Morgan 1990). Benzodiazepines were initially promoted as having a low dependence liability in comparison with barbiturates (Morgan 1990). However, by the 1970s prescribing had increased exponentially, and diazepam was the most widely prescribed drug in Europe and the USA (Licata and Rowlett 2008). Benzodiazepine dependence was identified to be a significant clinical concern (Tone 2005) In the 1980s, newer 'benzodiazepine-like' drugs including zolpidem and zopiclone were released, with the intention that they would have lower abuse liability, and shorter onset of action and duration making them ideal for insomnia. However, these newer drugs are also associated with abuse and dependence (Victorri-Vigneau et al. 2014).

2 Pharmacology and Clinical Effects

Benzodiazepines act through allosteric modulation of the $GABA_A$ receptor, increasing the affinity of the receptor for GABA leading to increased subsequent chloride conductance (Campo-Soria et al. 2006). Through their action at the $GABA_A$ receptor, benzodiazepines and benzodiazepine-like drugs enhance the activity of GABA, an inhibitory neurotransmitter, resulting in a slowing of neurotransmission and sedative and anxiolytic effects.

Most therapeutic uses of benzodiazepines are related to their anxiolytic, muscle relaxant and sleep-promoting effects. The effects of therapeutic doses of benzodiazepines on physiological parameters including respiration have been examined in non-opioid-maintained subjects. Administration of therapeutic doses of diazepam to healthy controls has been demonstrated to induce significant impairment of mental alertness and cognitive performance without producing significant effects on respiration (Mak et al. 1993; Bond 1993). Benzodiazepine has considerable effects on cognition, with low doses being sufficient to significantly impair driving ability (Verster et al. 2002). Benzodiazepines have been demonstrated to have an effect on memory, specifically newly learned material following benzodiazepine

Table 1 Benzodiazepine half-lives and dose equivalence

Benzodiazepine with examples of common brand names	Half-life	Dose equivalent to diazepam 5 mg (Therapeutic guidelines limited 2015) (mg)
Alprazolam: Kalma®, Xanax®	Short–intermediate	0.5
Bromazepam: Lexotan®	Short–intermediate	3
Clobazam: Frisium®	Intermediate	10
Clonazepam: Rivotri®, Paxam®	Intermediate	0.25
Diazepam: Antenex®, Ducene®, Valium®	Long	5
Flunitrazepam: Hypnodorm®, Rohypnol®	Intermediate	0.5
Lorazepam: Ativan®	Short–intermediate	1
Nitrazepam: Alodorm®, Mogadon®	Intermediate	5
Oxazepam: Alepam®, Murelax®, Serepax®	Short	15
Temazepam: Euhypnos®, Normison®, Temaze®	Short	10

administration (Curran 1986; Verster and Volkerts 2004). These amnestic effects can be desirable during surgical procedure.

Benzodiazepines are often classified as short, medium (or intermediate), or long acting depending on their duration of action and the time it takes for them to be cleared from the body. For example, midazolam is a short-acting benzodiazepine with an onset of 2–10 min depending on the route of administration and short duration of action lasting up to a few hours (Therapeutic Guidelines Limited 2015). In contrast, diazepam is a long-acting benzodiazepine with a half-life of 24–36 h (see Table 1).

3 Abuse and Dependence Liability

Midazolam, triazolam, flunitrazepam, and diazepam have been shown to demonstrate dose-dependent reinforcing effects in animals (Bai et al. 2011; Fischer and Rowlett 2011; Gomez et al. 2002; Gerak et al. 2001) abuse potential in humans (Carter 2007).

Certain characteristics are associated with greater self-administration of benzodiazepines under experimental conditions. Amongst healthy adolescents, benzodiazepine effects were found to vary as a function of sensation seeking, where those that scored higher on sensation-seeking scales reported greater sedative effects of diazepam, while low sensation seekers reported lower ratings measures indicative of abuse potential (Kelly et al. 2009). Diazepam has also found to be more reinforcing for people with social anxiety, and under experimental conditions that create anxiety (Helmus et al. 2005), and more reinforcing than buspirone in moderate alcohol consumers (Evans et al. 1996). Similarly, levels of alprazolam self-administration under double-blind conditions were positively associated with anxiety levels in patients with anxiety (Oswald et al. 1999). In contrast, amongst cannabis users, triazolam was not shown to function as a reinforcer at the doses examined, although significant effects were seen on participants' ratings of wanting to take the drug again and willingness to pay for the drug (Lile et al. 2010), which are generally associated with abuse liability. Taken together, these studies suggest that differences in characteristics and substance use may influence the likelihood of benzodiazepine use.

Despite their ability to maintain self-administration, benzodiazepines tend to have lower abuse liability compared with other drug classes such as opioids, barbiturates, cocaine, and GHB (Licata and Rowlett 2008; Carter et al. 2006). Further, drugs with shorter half-lives and shorter onset of action may have greater reinforcing effects (Licata and Rowlett 2008). Small studies suggest, for example, that alprazolam may have greater abuse liability than diazepam (Apelt et al. 1990), though few large rigorous study designs have examined this question. Long-term use can lead to neuroadaptation and physical dependence, further increasing the likelihood of abuse. Despite their lower propensity for self-administration compared to other drugs, misuse and the development of dependence are commonly reported amongst patient populations, either alone or in combination with other drugs. Dependence to benzodiazepines can begin to develop in as little as a week (Licata and Rowlett 2008), though not all patients develop dependence to benzodiazepines with long-term use (Woods et al. 1992).

Benzodiazepine dependence commonly presents with other drug dependence; benzodiazepine dependence as the primary drug of concern typically accounts for a very small proportion of treatment admission (e.g. less than one per cent in the USA) (Substance Abuse and Mental Health Services Administration 2011). Despite low numbers of treatment admissions, a large number of emergency data visits are now attributed to benzodiazepine use, with the presence of benzodiazepine use predicting a more serious outcome from an emergency department visit when present alone or with other drugs (Substance Abuse and Mental Health Services Administration 2014).

4 Indications for Benzodiazepines

One of the challenges with examining 'misuse' and 'abuse' of benzodiazepines is the large overlap between clinical use, self-administration for therapeutic purposes, and non-medical use. Non-prescribed use is commonly reported to be for reasons that appear therapeutic.

There are a range of reasons why benzodiazepines may be prescribed to patients, though benzodiazepines are rarely recommended as first-line treatments. Non-drug treatments and other medications such as antidepressants are considered first-line treatments for chronic anxiety or insomnia, with benzodiazepines reserved for second-line use when patients are unable to tolerate first-line medications, or after non-drug treatments have failed (Dellemijn and Fields 1994). Although benzodiazepines are effective when used acutely for generalized anxiety or panic disorders, they are not listed in clinical guidelines as first-line treatments for these conditions. For example, Australian guidelines indicate short-term use, or only where antidepressants are not tolerated (Therapeutic Guidelines Limited 2015; Joint Formulary Committee 2013). A recent international review of current guidelines report described the current role of benzodiazepines in generalized anxiety disorder to be largely limited to a second-line treatment for the acute phase, either until antidepressants or psychological treatments can be established, with caution using in specific populations including young people, the elderly, and those with a history of substance use disorder (Short- and Long-Term Use of Benzodiazepines in Patients with Generalized Anxiety Disorder 2014).

Benzodiazepines are also commonly used amongst those prescribed opioids for chronic pain (Nielsen et al. 2015). One review, conducted two decades ago, identified a limited role for benzodiazepines in acute pain in only a small number of conditions with little evidence from controlled studies to support their general use in chronic pain (Dellemijn and Fields 1994). Despite this, around one in three chronic pain patients continue to be prescribed benzodiazepines (Nielsen et al. 2015).

5 Prevalence of Use and Misuse

In the USA, national household surveys reveal that around 4 % of respondents reported tranquilizer use and 6 % reported using sleeping pills or other sedative use (Brower et al. 2011). Similar estimates from the UK report 3 % of the population use benzodiazepines (Ohayon et al. 1998). In Australia, over 5 million prescriptions for benzodiazepines are subsidized by the government each year (Medicare Australia 2011), accounting for approximately 4–5 % of all prescriptions written by general practitioners (Johnson et al. 2007). Many more benzodiazepines are supplied as private prescriptions that are not captured in any routine monitoring systems. Although there has been some reduction in benzodiazepine use (Islam et al. 2014; Tsimtsiou et al. 2009), they continue to be commonly prescribed, despite few indications for their use existing.

In some clinical populations, benzodiazepine use far exceeds that seen in the general population. For example, patients taking long-term opioids, both for chronic pain and in the context of treatment for illicit opioid use, have much higher rates of benzodiazepine use than the general populations (Nielsen et al. 2015; Ross and Darke 2000). Preclinical studies suggest that benzodiazepine may modulate the rewarding effects of heroin (Walker and Ettenberg 2001). Consistent with this, human laboratory-based studies where opioids and benzodiazepines are coadministered indicate that there may be additive subjective effects of opioids and benzodiazepines, with benzodiazepine administration potentially increasing the subjective opioid effects of methadone (Preston et al. 1984). Alcohol-dependent populations may also have higher rates of benzodiazepine use than the general public (Ciraulo et al. 1988), with benzodiazepines being commonly used to treat alcohol withdrawal. Alcohol-dependent people also have high rates of anxiety disorders which may increase susceptibility to benzodiazepine misuse (Helmus et al. 2005). High rates of benzodiazepine use have been reported amongst nursing home populations, which is concerning in light of increased fall risk that has been attributed to benzodiazepines (de Vries et al. 2013). One of the complicating characteristics of chronic benzodiazepine use is that when long-term use leads to dependence, the withdrawal symptoms (such as insomnia and anxiety) are similar to the initial symptoms the drugs are used to treat (Charney et al. 2006). This can result in patients perceiving that they are still treating their incident condition rather than having developed dependence.

Limited detailed research examines benzodiazepine use at a population level. A recent US study on benzodiazepine use amongst adolescents used data from the National Survey of Drug Use and Health (NSDUH) to examine characteristics and theoretical risk factors associated with benzodiazepine use amongst adolescents (Rigg and Ford 2014). The authors found that being female, being older, and viewing substances use as less risky were associated with lifetime benzodiazepine misuse. Those with peers with more lenient attitudes towards substance use and who were under more strain (based on an index of negative life events such as arguments with parents or low grades in school) were also associated with increased risk of misuse. Other substances use was also associated with increased risk, consistent with previous research that generally finds benzodiazepines are used in a context of polydrug use.

In a study of high school seniors (modal age of 18 years), approximately 5 % report medical use and 8 % report non-medical use of benzodiazepines (McCabe and West 2014). Correlates of non-medical use were similar in these two studies with being female and white being associated with increased non-medical use in both studies, along with use of other substances, including non-medical use of other types of pharmaceuticals.

Most research on benzodiazepine use at a population level comes from countries such as the UK and the USA, though misuse and dependence not limited to these countries. For example, in France, benzodiazepines are thought to be misused more often than most opioid analgesics, excluding morphine (Pauly et al. 2012). A household survey in Thailand identified around 4 % of the population were current benzodiazepine users, and 57 % of those that were using benzodiazepines reported misuse of them (Puangkot et al. 2011). Other studies identify considerable concerns, with emerging awareness of benzodiazepine-related problems reported in countries including Albania, India, and Lebanon (Kellici et al. 2013; Nattala et al. 2014; Naja et al. 2000).

5.1 *Adverse Effects from Benzodiazepine Use*

5.1.1 Dependence

Iatrogenic dependence is common with benzodiazepines due to the dependence liability of this class of drugs (Denis et al. 2006) and may occur following therapeutic use of benzodiazepines for a range of psychiatric conditions including anxiety and panic disorder as well as for the treatment of insomnia.

5.1.2 Aggression

Increased hostility, anger, and aggression have been reported following benzodiazepine administration in preclinical (Miczek et al. 1993) and human studies (DiMascio 1970). More than fifty years ago, reports of aggression were recorded with chlordiazepoxide (Boyle and Tobin 1961). Other early studies described rage attacks and 'egodystonic hatefulness' in association with diazepam use (Feldman 1962). Other unwanted benzodiazepine effects include disinhibition, paradoxical hostility, and anterograde amnesia (Daderman and Lidberg 1999; Dobbin 2001; Rall 1992; Bonn and Bonn 1998). Benzodiazepine use has been linked to criminal behaviour and in at-risk populations may contribute to further harms for both the user and the community (Jones et al. 2011). The finding of a dose-related effect of flunitrazepam on risky decision-making may partially explain these findings (Lane et al. 2007). Paradoxical hostility is an unexpected side effect of benzodiazepines, given their known ability to generally cause sedation and reduce anxiety. Daderman and Lidberg (1999) studied five forensic patients that demonstrated paradoxical reactions to flunitrazepam when it was used in combination with alcohol and other drugs (Daderman and Lidberg 1999). The reactions included hostility and anterograde amnesia, which were noted to be in contrast to patients based usual psychological characteristics. Similar paradoxical responses have been reported with benzodiazepines, including alprazolam and diazepam (French 1989; Rudorfer et al. 1989).

The mechanisms of paradoxical reactions are yet to be fully elucidated (Mancuso et al. 2004; Robin and Trieger 2002). Paradoxical reactions are thought to be more common amongst certain group of patients, including children, and those that use substances including alcohol. A possible role of genetics and serotonergic mechanism have been suggested. Common treatment approaches include administration of flumazenil, a benzodiazepine antagonist.

5.1.3 Dementia and Other Cognitive Effects

Long-term benzodiazepine use can result in cognitive impairment across numerous cognitive domains (Barker et al. 2004). A review of benzodiazepine use and dementia identified that 9 out of 10 studies examined identified an association, with increased risk of dementia following benzodiazepine use. When only the high-quality studies were examined, the risk was found to be increased by a factor of 1.24–2.30. The greatest risk was with higher doses, longer term use, and long half-life benzodiazepines. Use of longer than three years was associated with the greatest risk that did not disappear on cessation of benzodiazepines (Billioti de Gage et al. 2015).

5.1.4 Mortality

Clearly, the most concerning adverse effect associated with benzodiazepines is mortality. Benzodiazepine use is commonly implicated in opioid deaths, including heroin, methadone, and other prescription opioids (Zador and Sunjic 2000; Gerostamoulos et al. 2001; Caplehorn and Drummer 2002; Ernst et al. 2002; Jann et al. 2014). Similarly, benzodiazepine use has been implicated as a significant risk factor for non-fatal heroin overdose (Gutierrez-Cebollada et al. 1994; Neale 2000; Dietze et al. 2005).

Concerns with mortality exist in therapeutic as well as non-medical use. For example, a Swedish study of 2249 patients starting long-term oxygen therapy for COPD between 2005 and 2009 found that benzodiazepines were dose dependently associated with increased mortality (Ekstrom et al. 2014).

5.1.5 Falls

An association with benzodiazepine use and falls has been established. For example, an Irish cohort study of 6666 adults aged 50 years or more found that benzodiazepines were associated with a greater number of falls (Richardson et al. 2015). Further studies have tried to identify which benzodiazepine characteristics are associated with more falls; however, findings have been mixed. One prospective observational study found an association with only short-acting benzodiazepines and falls, with non-significant associations with long-acting benzodiazepines (de Vries et al. 2013). A separate matched case-control study also identified an association with shorter elimination half-life benzodiazepines, but found recent dose escalation and total dose of benzodiazepines to be more important contributors to falls (Herings et al. 1995).

6 The Spectrum of Benzodiazepine Use: From Therapeutic to Non-medical Use and Dependence

Optimal use of benzodiazepines is outlined in clinical guidelines (Practitioners RRACoG 2000). General feature of optimal use includes the following:

- Avoiding prescribing in high-risk groups (e.g. people with substance use disorder);
- Advising patients of the risk of dependence;
- Having only one prescriber providing prescriptions;
- Regular review;
- Use of non-drug management for conditions such as anxiety and insomnia; and
- Using the lowest dose for the shortest period possible (no longer than 2–4 weeks).

Where dependence has been established, with the patients consent, it is recommended to transfer doses to one long-acting benzodiazepine and gradually reduce (often over several months).

Unfortunately, much benzodiazepine use occurs outside these parameters with benzodiazepines commonly being used for long periods of time in the absence of other strategies to treat the primary indication. Further complicating this clinical picture is the similarity between benzodiazepine withdrawal symptoms (e.g. agitation, anxiety, insomnia, and panic attacks) and the indications they are initially commenced for.

Box 1. A case of escalating benzodiazepine use Rachel was a final-year nursing student under a lot of financial stress and approaching her examinations. She was finding it increasingly difficult to sleep at night and went to see her doctor for help. She received an initial script for temazepam 10 mg (25 tablets) and was advised just to take it for a few days. She found the temazepam helped and after a few days attempted to sleep again without them. Her insomnia was still a problem, and Rachel decided to keep taking them just until she was finished with her examinations. Three weeks later, in the middle of her examinations, she ran out of temazepam and had 'the worst night of sleep ever'. She returned for another script of temazepam just to get her through the last few days of examinations. She ended up finishing the second bottle, and by this stage, she had been taking benzodiazepines to sleep almost every day for two months. She was finding that they were not working as well, and she needed a higher dose to get the same effect. After a few months, if she went without them, in addition to having a lot of trouble sleeping, she would experience anxiety and panic attacks during the day. After pleading with her doctor for something to help her sleep and to help with the panic attacks, she was prescribed a higher dose of temazepam and alprazolam for panic attacks during the day. After continuing on these medications, a number of months Rachel decided to seek help for her panic attacks as they were getting worse. She disclosed at this time that her benzodiazepine use was now much higher than intended and that she was seeing multiple doctors to make sure she did not run out. With the help of her family doctor and a specialist in addiction medicine, she was eventually converted onto a long-acting benzodiazepine (diazepam) started on a withdrawal programme, where she reduced over 6 months and eventually stopped all benzodiazepines. She worked with a psychologist to learn cognitive behavioural techniques to help with her sleep and panic attacks. She was still experiencing symptoms of anxiety and insomnia several months after ceasing benzodiazepines.

Some of the challenges in addressing benzodiazepine 'misuse' is that their use is often in the context of self-medication of insomnia and other psychological disturbances [e.g. (Gelkopf et al. 1999; Perera et al. 1987)]. In other cases, benzodiazepine use occurs in the context of polydrug use, for example with opioids to either increase opioid effects, for example, in times of lower purity illicit opioids or where greater euphoric effects are desired [e.g. (Darke et al. 1995; Iguchi et al. 1993)].

A large study of 1048 patients that had received benzodiazepines on prescription for at least one month examined characteristics and risk factors associated with dependence (de las Cuevas et al. 2003). Unsurprisingly, increasing dose and duration was correlated with the rate of developing dependence. The third factor associated with dependence was coprescription of an antidepressant, potentially acting as a proxy for poorer psychological health, though this was not significant after controlling for dose and duration of benzodiazepines.

Amongst older adults (65 years and older), prescribed benzodiazepines for at least three months, around three in ten, were identified to be dependent (Yen et al. 2014). Use of zolpidem (as opposed to estazolam or flunitrazepam) and current depression were associated with misuse (Yen et al. 2014). A separate study of older adults in Quebec identified that while one in ten met diagnostic criteria for substance dependence, almost half identified as being dependent and a third agreed it would be good to stop taking benzodiazepines (Voyer et al. 2010). A third study examined outcomes of benzodiazepine prescribing to older adults who were newly initiated onto benzodiazepines (Simon and Ludman 2006). This study identified that insomnia (42 %) and anxiety (36 %) were the most common reasons for benzodiazepine prescription and that 30 % of those initiated on benzodiazepines were taking them daily after two months.

Recreational use of benzodiazepines is not uncommon. Amongst a small sample ($n = 15$) of college students, the most common reasons for misuse were 'to get high or party' (33 %), or to relax or 'zone out' (27 %) (Stone and Merlo 2011).

A recent survey in the UK identified that around three out of ten people who have taken benzodiazepines had 'misused' them. Some of the reasons most commonly reported for misuse were largely similar to their therapeutic indications such as sleep (66 %) and help cope with stress (37 %), or for recreational reasons such as 'to get high' (31 %) or for social reasons (24 %) (Kapil et al. 2014). One in ten reported that they misused them because they felt they were safer than street/illegal drugs (Kapil et al. 2014).

7 Benzodiazepines and Polydrug Use

Benzodiazepine use amongst methadone patients has also been found to be common in many settings, with recent use reported by 44–70 % and lifetime use reported by up to 100 % amongst different methadone treatment samples (Gelkopf et al. 1999; Iguchi et al. 1993; Stitzer et al. 1981; Barnas et al. 1992; Hartog and Tusel 1987; Chen et al. 2011). Similarly, benzodiazepine use has been reported to be common amongst buprenorphine treatment participants (Nielsen et al. 2007).

In- and out-of-treatment samples of people who use heroin commonly report benzodiazepine use; around 25 % of heroin users are believed to be benzodiazepine dependent (Ross and Darke 2000; Darke et al. 1992). Amongst Australian heroin-dependent treatment entrants, just over half (52 %) reported using benzodiazepines in the previous month (Ross et al. 2005).

Different characteristics have been reported amongst patients seeking treatment for benzodiazepine abuse or dependence. One study of $n = 176$ patients referred for assessment and treatment of their benzodiazepine abuse and/or dependence found at least two subgroups treatment seeking of benzodiazepine users (Busto et al. 1986). The first was a group using only benzodiazepines in lower diazepam equivalent doses (approximately 15 mg daily), while a second group was using multiple substances and higher benzodiazepine doses, (approximately 40 mg daily of diazepam equivalents) with more dose escalation. A larger study of 2440 people receiving long-term benzodiazepines had similar findings: most recipients of a long-term benzodiazepine script did not increase their dose and remained on a low dose (around 10 mg daily of diazepam equivalents), though a small subset (1.6 %) did increase their dose (Soumerai et al. 2003). This subgroup were characterized by concurrent antidepressant use, use of multiple pharmacies, and younger age.

8 Perceptions of Benzodiazepine Use

Amongst people who use drugs, benzodiazepine use is often perceived to be less of a concern than other drugs. Amongst parents of young children participating in opioid treatment, benzodiazepine use and benzodiazepine dependence were relatively normalized, with reducing use reported to be less of a priority than addressing opioid use (Chandler et al. 2014). While opioid use was perceived as stigmatizing, benzodiazepine use was perceived to be a helpful part of their 'normal' life (Chandler et al. 2014). Reasons reported for use included insomnia and helping with nerves and mood with few adverse effects noted. A separate qualitative study of polydrug users reported similarly that benzodiazepines were perceived to be 'less risky' than illicit drugs such as heroin due to being of a known quantity (Fountain et al. 1999).

Patient perceptions often also suggest benzodiazepine use is in the context of self-medication. A qualitative study of opioid treatment patients in Virginia reported that benzodiazepines are often used in combination with opioids for anxiety, commonly referring to them as 'nerve pills' (Redican et al. 2012).

Perceptions of older women, a population who are overrepresented amongst benzodiazepine-dependent people (Voyer et al. 2010), suggested that amongst some there was some confusion of what dependence is, drawing parallels with needing medication for diabetes and blood pressure medications, and drawing distinctions between physical dependence and 'addiction' (noting the latter as a concern and denying being 'addicted') (Canham et al. 2014). Participants also reported that use was not problematic as they had been 'told to' take their medications by a doctor, reserving the concept of 'addiction' for illicit drugs.

A study of young to middle-aged people ($n = 212$) taking long-term benzodiazepines found half (49.7 %) met diagnostic criteria for benzodiazepine dependence, though only a small group of these patients ($n = 29$) reported acquiring them in ways other than sanctioned medical supply (Guerlais et al. 2015). This study identified two clinical profiles of benzodiazepine-dependent people, one with tolerance (and increasing doses) and long-term use, and a second with concern about use and somatic consequences of benzodiazepine use. In this study, the greater benzodiazepine use was associated with substance use and psychiatric disorders.

9 Responses to Benzodiazepine Misuse and Dependence

Given the considerable harms associated with benzodiazepine use, there are relatively few widely used clinical responses.

Prescription monitoring is commonly proposed to address pharmaceutical misuse. Where prescription monitoring has been implemented, significant reductions in inappropriate benzodiazepine use have been reported (Gomes et al. 2014). It should be noted that, as a result of benzodiazepines being classed in different levels of regulation than strong prescription opioids, they are sometimes excluded from prescription drug monitoring programmes (Nielsen 2014).

Lower levels of reimbursement have also been associated with reduced benzodiazepine use (Hoebert et al. 2012).

Strategies to monitor and respond to benzodiazepine use in clinical care include routine urine drug screen, to detect non-prescribed use (Gudin et al. 2013). Where low doses of benzodiazepines are prescribed, the use of benzodiazepines such as clonazepam and clobazam that do not have active metabolites can help aid in the detection of unsanctioned benzodiazepine use (Lintzeris and Nielsen 2010).

Two systematic reviews address strategies for reducing the use of benzodiazepines. A Cochrane review that examined data from with 458 subjects across eight randomized controlled trials (RCTs) (Denis et al. 2006) identified supported for transferring patients to a long-acting benzodiazepines, stabilization preceding a gradual taper over weeks to months, and found a limited role for adjuvant medication therapy (including beta blockers, some tricyclic antidepressants such as dothiepin, buspirone, and progesterone). Carbamazepine showed modest effects in reducing benzodiazepine withdrawal symptoms (Schweizer et al. 1991).

A second review (Parr et al. 2009) identified several RCTs, most conducted one to two decades ago, using adjuvant medications. Results included inconsistent findings with paroxetine and positive findings in one study for trazodone and sodium valproate. It should be noted that most controlled studies in this area were conducted 1–2 decades ago, and the role of newer medications in assisting benzodiazepines withdrawal has not yet been explored in controlled trials. Promising anecdotal and case reports regarding the role of new-generation antidepressants (e.g. mirtazapine), new-generation antipsychotic medications (e.g. olanzapine, aripiprazole), medications that impact upon GABA receptor system (e.g. gabapentin), and benzodiazepine receptor antagonists (e.g. flumazenil) warrant further research.

Parr et al. (2009) also examined the role of psychosocial interventions in addition to gradual dose reduction (Parr et al. 2009). Common features of these interventions included relaxation training, symptom management, and cognitive behavioural techniques. Findings were variable, with positive outcomes from psychosocial interventions in four of the eight studies.

While research identified approaches to reduce benzodiazepine doses, many patients appear unable to maintain long-term abstinence from benzodiazepine dependence. For example, benzodiazepine in a general practice setting with older adults (mean age 63 years) found that despite using relatively low doses of benzodiazepines at baseline (mean diazepam equivalent dose of 8.4 mg), most (88 %) had recommenced benzodiazepine use within 15 months of benzodiazepine reduction (Oude Voshaar et al. 2006). The difficulties experienced in tapering patients off benzodiazepines highlight the importance of strategies that prevent the development of dependence, specifically, addressing inappropriate prescribing that leads to dependence. This is in addition to identifying effective medications to assist in withdrawal for those that do develop dependence.

References

Apelt S, Schmauss C, Emrich HM (1990) Preference for alprazolam as opposed to diazepam in benzodiazepine-dependent psychiatric inpatients. Pharmacopsychiatry 23(2):70–5 (PubMed PMID: 1971118. Epub 1990/03/01. eng)

Bai X, France CP, Gerak LR (2011) The discriminative stimulus effects of midazolam are resistant to modulation by morphine, amphetamine, dizocilpine, and gamma-butyrolactone in rhesus monkeys. Psychopharmacol (Berl) 217(4):495–504 (PubMed PMID: 21503606. Pubmed Central PMCID: PMC3195358. Epub 2011/04/20. eng)

Barker MJ, Greenwood KM, Jackson M, Crowe SF (2004) Cognitive effects of long-term benzodiazepine use: a meta-analysis. CNS Drugs 18(1):37–48

Barnas C, Rossmann M, Roessler H, Riemer Y, Fleischhacker WW (1992) Benzodiazepines and other psychotropic drugs abused by patients in a methadone maintenance program: familiarity and preference. J Clin Psychopharmacol 12(6):397–402

Billioti de Gage S, Pariente A, Begaud B (2015) Is there really a link between benzodiazepine use and the risk of dementia? Expert Opinion Drug Safety 14(5):733–747 (PubMed PMID: 25691075. Epub 2015/02/19. eng)

Bond A (1993) The risks of taking benzodiazepines. In: Hallstrom C (ed) Benzodiazepines dependence, 1st edn. Oxford University Press, London, pp 34–45

Bonn D, Bonn J (1998) Anxious times for the treatment of anxiety. Lancet 352(3):1126

Boyle D, Tobin J (1961) Pharmaceutical management of behaviour disorders: chlordiazepoxide in covert and overt expressions of aggression. J Med Soc NJ 58:427–429

Brower KJ, McCammon RJ, Wojnar M, Ilgen MA, Wojnar J, Valenstein M (2011) Prescription sleeping pills, insomnia, and suicidality in the national comorbidity survey replication. J Clin Psychiatry 72(4):515–521

Busto U, Sellers EM, Naranjo CA, Cappell HD, Sanchez-Craig M, Simpkins J (1986) Patterns of benzodiazepine abuse and dependence. Br J Addict 81(1):87–94 (PubMed PMID: 2870731. Epub 1986/02/01. eng)

Campo-Soria C, Chang Y, Weiss DS (2006) Mechanism of action of benzodiazepines on GABAA receptors. Br J Pharmacol 148(7):984–90 (PubMed PMID: 16783415. Pubmed Central PMCID: PMC1751932. Epub 2006/06/20. eng)

Canham SL, Gallo J, Simoni-Wastila L (2014) Perceptions of benzodiazepine dependence among women age 65 and older. J Gerontol Soc Work 57(8):872–888

Caplehorn JR, Drummer OH (2002) Fatal methadone toxicity: signs and circumstances, and the role of benzodiazepines. Aust NZ J Publ Health 26(4):358–62

Carter LP, Richards BD, Mintzer MZ, Griffiths RR (2006) Relative abuse liability of GHB in humans: a comparison of psychomotor, subjective, and cognitive effects of supratherapeutic doses of triazolam, pentobarbital, and GHB. Neuropsychopharmacol 31(11):2537–2551 (PubMed PMID: 16880774. Epub 2006/08/02. eng)

Carter LP, Griffiths RR, Suess PE, Casada JH, Wallace CL, Roache JD (2007) Relative abuse liability of indiplon and triazolam in humans: a comparison of psychomotor, subjective, and cognitive effects. J Pharmacol Exp Ther 322(2):749–59 (PubMed PMID: 17502431. Epub 2007/05/16. eng)

Chandler A, Whittaker A, Williams N, McGorm K, Cunningham-Burley S, Mathews G (2014) Mother's little helper? Contrasting accounts of benzodiazepine and methadone use among drug-dependent parents in the UK. Drugs (Abingdon, England) 21(6):470–5 (PubMed PMID: 25552821. Pubmed Central PMCID: PMC4266080. Epub 2015/01/02. eng)

Charney DS, Mihic SJ, Harris RA (2006) Hypnotics and sedatives. In: Laurence L, Brunton P (eds) Goodman and Gilman's the pharmacological basis of therapeutics. Printed in the United States of America: Mcgraw-hill, Medical Publishing Division

Chen KW, Berger CC, Forde DP, D'Adamo C, Weintraub E, Gandhi D (2011) Benzodiazepine use and misuse among patients in a methadone program. BMC psychiatry 11:90 (PubMed PMID: 21595945. Pubmed Central PMCID: PMC3117775. Epub 2011/05/21. eng)

Ciraulo DA, Sands BF, Shader RI (1988) Critical review of liability for benzodiazepine abuse among alcoholics. Am J Psychiatry 145(12):1501–1506 (PubMed PMID: 2904227. Epub 1988/12/01. eng)

Curran HV (1986) Tranquillising memories: A review of the effects of benzodiazepines on human memory. Biol Psychol 23(2):179–213

Daderman AM, Lidberg L (1999) Flunitrazepam (Rohypnol) abuse in combination with alcohol causes premeditated, grievous violence in male juvenile offenders. J Am Acad Psychiatry Law 27(1):83–99

Darke S, Hall W, Ross M, Wodak A (1992) Benzodiazepine use and HIV risk-taking behaviour among injecting drug users. Drug Alcohol Depend 1992(31):31–36

Darke SG, Ross JE, Hall WD (1995) Benzodiazepine use among injecting heroin users. Med J Aust 162(12):645–647

de las Cuevas C, Sanz E, de la Fuente J (2003) Benzodiazepines: more "behavioural" addiction than dependence. Psychopharmacology (Berl) 167(3):297–303 (PubMed PMID: 12669174. Epub 2003/04/02. eng)

de Vries OJ, Peeters G, Elders P, Sonnenberg C, Muller M, Deeg DJ, et al (2013) The elimination half-life of benzodiazepines and fall risk: two prospective observational studies. Age Ageing 42(6):764–70 (PubMed PMID: 23900130. Epub 2013/08/01. eng)
Dellemijn PLI, Fields HL (1994) Do benzodiazepines have a role in chronic pain management? Pain 57(2):137–52
Denis C, Auriacombe M, Fatsas M, Lavie E (2006) Pharmacological interventions for benzodiazepine mono-dependence management in outpatient settings. Cochrane Database Syst Rev 3(2006):CD005194
Dietze P, Jolley D, Fry C, Bammer G (2005) Transient changes in behaviour lead to heroin overdose: results from a case-crossover study of non-fatal overdose. Addiction 100(5):636–642
DiMascio A (1970) The effects of benzodiazepines on aggression: Reduced or increased? Psychopharmacology 30:95–102
Dobbin M (2001) Discussion paper: temazepam injecting in Victoria 2001. Drugs Policy and Services, Victorian Department of Human Services, Melbourne
Ekstrom MP, Bornefalk-Hermansson A, Abernethy AP, Currow DC (2014) Safety of benzodiazepines and opioids in very severe respiratory disease: national prospective study. BMJ 348: g445 (PubMed PMID: 24482539. Pubmed Central PMCID: PMC3906915. Epub 2014/02/01. eng)
Ernst E, Bartu A, Popescu A, Ileutt KF, Hansson R, Plumley N (2002) Methadone-related deaths in Western Australia 1993–99. Aust NZ J Publ Health 26(4):364–370
Evans SM, Griffiths RR, de Wit H (1996) Preference for diazepam, but not buspirone, in moderate drinkers. Psychopharmacol (Berl) 123(2):154–163 (PubMed PMID: 8741938. Epub 1996/01/01. eng)
Feldman PE (1962) An analysis of the efficacy of diazepam. J Neuropsychiatry 3:62–67
Fischer BD, Rowlett JK (2011) Anticonflict and reinforcing effects of triazolam + pregnanolone combinations in rhesus monkeys. J Pharmacol Exp Ther 337(3):805–11 (PubMed PMID: 21411495. Pubmed Central PMCID: PMC3101007. Epub 2011/03/18. eng)
Fountain J, Griffiths P, Farrell M, Gossop M, Strang J (1999) Benzodiazepines in polydrug-using repertoires: the impact of the decreased availability of temazepam gel-filled capsules. Drugs Edu Prev Policy 6(1)
French AP (1989) Dangerously aggressive behavior as a side effect of alprazolam. Am J Psychiatry 146(2):276
Gelkopf M, Bleich A, Hayward R, Bodner G, Adelson M (1999) Characteristics of benzodiazepine abuse in methadone maintenance treatment patients: a 1 year prospective study in an Israeli clinic. Drug Alcohol Depend 55(1–2):63–68
Gerak LR, Woolverton WL, Nader MA, Patrick GA, Harris LS, Winger G, et al (2001) Behavioral effects of flunitrazepam: reinforcing and discriminative stimulus effects in rhesus monkeys and prevention of withdrawal signs in pentobarbital-dependent rats. Drug Alcohol Depend 63 (1):39–49 (PubMed PMID: 11297830. Epub 2001/04/12. eng)
Gerostamoulos J, Staikos V, Drummer OH (2001) Heroin-related deaths in Victoria: a review of cases for 1997 and 1998. Drug Alcohol Depend 61(2):123–127
Gomes T, Juurlink D, Yao Z, Camacho X, Paterson JM, Singh S, et al (2014) Impact of legislation and a prescription monitoring program on the prevalence of potentially inappropriate prescriptions for monitored drugs in Ontario: a time series analysis. CMAJ open 2(4):E256–61 (PubMed PMID: 25485251. Pubmed Central PMCID: PMC4251507. Epub 2014/12/09. eng)
Gomez TH, Roache JD, Meisch RA (2002) Relative reinforcing effects of different benzodiazepine doses for rhesus monkeys. Drug Alcohol Depend 68(3):275–83 (PubMed PMID: 12393222. Epub 2002/10/24. eng)
Gudin JA, Mogali S, Jones JD, Comer SD (2013) Risks, management, and monitoring of combination opioid, benzodiazepines, and/or alcohol use. Postgrad Med 125(4):115–130 (PubMed PMID: 23933900. Pubmed Central PMCID: PMC4057040. Epub 2013/08/13. eng)
Guerlais M, Grall-Bronnec M, Feuillet F, Gérardin M, Jolliet P, Victorri-Vigneau C (2015) Dependence on prescription benzodiazepines and z-drugs among young to middle-aged patients in France. Subst Use Misuse 50(3):320–327

Gutierrez-Cebollada J, de la Torre R, Ortuno J, Garces JM, Cami J (1994) Psychotropic drug consumption and other factors associated with heroin overdose. Drug Alcohol Depend 35 (2):169–174

Hartog J, Tusel DJ (1987) Valium use and abuse by methadone maintenance clients. Int J Addict 22(11):1147–1154

Helmus TC, Tancer M, Johanson CE (2005) Reinforcing effects of diazepam under anxiogenic conditions in individuals with social anxiety. Exp Clin Psychopharmacol 13(4):348–56 (PubMed PMID: 16366765. Epub 2005/12/22. eng)

Herings RM, Stricker BH, de Boer A, Bakker A, Sturmans F (1995) Benzodiazepines and the risk of falling leading to femur fractures. Dosage more important than elimination half-life. Arch Intern Med 155(16):1801–1807 (PubMed PMID: 7654115. Epub 1995/09/11. eng)

Hoebert JM, Souverein PC, Mantel-Teeuwisse AK, Leufkens HG, van Dijk L (2012) Reimbursement restriction and moderate decrease in benzodiazepine use in general practice. Annals Fam Med 10(1):42–49 (PubMed PMID: 22230829. Pubmed Central PMCID: PMC3262472. Epub 2012/01/11. eng)

Iguchi MY, Handelsman L, Bickel WK, Griffiths RR (1993) Benzodiazepine and sedative use/abuse by methadone maintenance clients. Drug Alcohol Depend 32(3):257–266

Islam MM, Conigrave KM, Day CA, Nguyen Y, Haber PS (2014) Twenty-year trends in benzodiazepine dispensing in the Australian population. Intern Med J 44(1):57–64

Jann M, Kennedy WK, Lopez G (2014) Benzodiazepines: a major component in unintentional prescription drug overdoses with opioid analgesics. J Pharm Pract 27(1):5–16 (PubMed PMID: 24436437. Epub 2014/01/18. eng)

Johnson CBB, Brough R, Buchanan J (2007) Benzodiazepine prescribing: lessons from interprofessional dialogue. Aust Fam Physician 36(4):245–306)

Joint Formulary Committee (2013) British national formulary (BNF) 66. BMJ Publishing Group Ltd and Royal Pharmaceutical Society, London

Jones K, Nielsen S, Bruno R, Frei M, Lubman D (2011) Benzodiazepine prescribing and violence: implications for general practitioners. Aust Fam Phys 40(11):862–865

Kapil V, Green JL, Le Lait C, Wood DM, Dargan PI (2014) Misuse of benzodiazepines and Z-drugs in the UK. Br J Psychiatry 205(5):407–408 (PubMed PMID: 25061118. Epub 2014/07/26. eng)

Kellici S, Hoti E, Burazeri G (2013) Level and factors of benzodiazepines misuse in Albania. Int J Clin Pharm 35(3):323–326 (PubMed PMID: 23423641. Epub 2013/02/21. eng)

Kelly TH, Delzer TA, Martin CA, Harrington NG, Hays LR, Bardo MT (2009) Performance and subjective effects of diazepam and d-amphetamine in high and low sensation seekers. Behav Pharmacol 20(5–6):505–517 (PubMed PMID: 19654505. Pubmed Central PMCID: PMC3148198. Epub 2009/08/06. eng)

Lane SD, Cherek DR, Nouvion SO (2007) Modulation of human risky decision making by flunitrazepam. Psychopharmacology 196(2):177–188 (PubMed PMID: 17917718. Epub 2007/10/06. eng)

Licata SC, Rowlett JK (2008) Abuse and dependence liability of benzodiazepine-type drugs: GABA(A) receptor modulation and beyond. Pharmacol biochem behav 90(1):74–89 (PubMed PMID: 18295321. Pubmed Central PMCID: PMC2453238. Epub 2008/02/26. eng)

Lile JA, Kelly TH, Hays LR (2010) The reinforcing, self-reported, performance and physiological effects of Δ(9)-tetrahydrocannabinol, triazolam, hydromorphone and methylphenidate in cannabis users. Behav Pharmacol 21(1):29–38 (PubMed PMID: PMC2903043)

Lintzeris N, Nielsen S (2010) Benzodiazepines, methadone and buprenorphine: interactions and clinical management. Am J Addict 19(1):59–72

Mak KH, Wang YT, Cheong TH, Poh SC (1993) The effect of oral midazolam and diazepam on respiration in normal subjects. Eur Respir J 6(1):42–47

Mancuso CE, Tanzi MG, Gabay M (2004) Paradoxical reactions to benzodiazepines: literature review and treatment options. Pharmacotherapy 24(9):1177–1185 (PubMed PMID: 15460178. Epub 2004/10/06. eng)

McCabe SE, West BT (2014) Medical and nonmedical use of prescription benzodiazepine anxiolytics among U.S. high school seniors. Addict Behav 39(5):959–664 (PubMed PMID: 24556157. Pubmed Central PMCID: PMC4312492. Epub 2014/02/22. eng)

Medicare Australia (2011) https://www.medicareaustralia.gov.au/statistics/pbs_item.shtml [database on the Internet]. Australian Government 2011 [cited 5th July 2011]

Miczek KA, Weerts EM, Debold JF (1993) Alcohol, benzodiazepine (GABA sub A) receptor complex and aggression: Ethological analysis of individual differences in rodents and primates. J Stud Alcohol 11:170–179

Morgan WW (1990) Abuse liability of barbiturates and other sedative-hypnotics. Adv Alcohol Subst Abuse 9(1–2):67–82 (ubMed PMID: 2198786. Epub 1990/01/01. eng)

Naja WJ, Pelissolo A, Haddad RS, Baddoura R, Baddoura C (2000) A general population survey on patterns of benzodiazepine use and dependence in Lebanon. Acta Psychiatrica Scandinavica 102(6):429–431 (PubMed PMID: 11142431. Epub 2001/01/06. eng)

Nattala P, Murthy P, Thennarasu K, Cottler LB (2014) Nonmedical use of sedatives in urban Bengaluru. Indian J Psychiatry 56(3):246–252 (PubMed PMID: 25316935. Pubmed Central PMCID: PMC4181179. Epub 2014/10/16. eng)

Neale J (2000) Methadone, methadone treatment and non-fatal overdose. Drug Alcohol Depend 58 (1–2):117–124

Nielsen SBR (2014) Implementing real time prescription drug monitoring: are we ready? (Editorial). Drug Alcohol Rev. Accepted 14 July 2014 (In Press)

Nielsen S, Dietze P, Lee N, Dunlop A, Taylor D (2007) Concurrent buprenorphine and benzodiazepines use and self-reported opioid toxicity in opioid substitution treatment. Addiction (Abingdon, England) 102(4):616–622

Nielsen S, Lintzeris N, Bruno R, Campbell G, Larance B, Hall W et al (2015) Benzodiazepine use amongst chronic pain patients prescribed opioids: associations with pain, physical and mental health and health service utilization. Pain Med 16(2):356–366

Ohayon MM, Caulet M, Priest RG, Guilleminault C (1998) Psychotropic medication consumption patterns in the UK general population. J Clin Epidemiol 51(3):273–83

Oswald LM, Roache JD, Rhoades HM (1999) Predictors of individual differences in alprazolam self-medication. Exp Clin Psychopharmacol 7(4):379–90 (PubMed PMID: 10609973. Epub 1999/12/28. eng)

Oude Voshaar RC, Couvee JE, Van Balkom AJLM, Mulder PGH, Zitman FG (2006) Strategies for discontinuing long-term benzodiazepine use: meta-analysis. Br J Psychiatry 189(3):213–220

Parr JM, Kavanagh DJ, Cahill L, Mitchell G, Mc DYR (2009) Effectiveness of current treatment approaches for benzodiazepine discontinuation: a meta-analysis. Addiction 104(1):13–24 (PubMed PMID: 18983627. Epub 2008/11/06. eng)

Pauly V, Pradel V, Pourcel L, Nordmann S, Frauger E, Lapeyre-Mestre M, et al (2012) Estimated magnitude of diversion and abuse of opioids relative to benzodiazepines in France. Drug Alcohol Depend 126(1–2):13–20

Perera KM, Tulley M, Jenner FA (1987) The use of benzodiazepines among drug addicts. Br J Addict 82(5):511–5 (PubMed PMID: 2885020. Epub 1987/05/01. eng)

Practitioners RRACoG (2000) RACGP guidelines for rational use of benzodiazepines Melbourne. RACGP [updated October; cited 2011 27th June]

Preston KL, Griffiths RR, Stitzer ML, Bigelow GE, Liebson IA (1984) Diazepam and methadone interactions in methadone maintenance. Clin Pharmacol Ther 36(4):534–541 (PubMed PMID: 6478738. Epub 1984/10/01. eng)

Puangkot S, Laohasiriwong W, Saengsuwan J, Chiawiriyabunya I (2011) Prevalence of benzodiazepines misuse in Ubon Ratchathani province Thailand. J Med Assoc Thai = Chotmaihet thangphaet 94(1):118–22 (PubMed PMID: 21425737. Epub 2011/03/24. eng)

Rall T (1992) Hypnotics and sedatives: ethanol. In: Goodman Gilman ATW, Nies A, Taylor P (eds), Goodman and Gilman's the pharmacological basis of therapeutics. McGraw-Hill Inc, New York, p 345–382

Redican KJ, Marek LI, Brock DJ, McCance-Katz EF (2012) Exploring the etiologic factors and dynamics of prescription drug abuse in southwest virginia. Health Promot Perspect 2(2):153–165 (PubMed PMID: 24688929. Pubmed Central PMCID: PMC3963639. Epub 2012/01/01. eng)

Richardson K, Bennett K, Kenny RA (2015) Polypharmacy including falls risk-increasing medications and subsequent falls in community-dwelling middle-aged and older adults. Age Ageing 44(1):90–6 (PubMed PMID: 25313240. Epub 2014/10/15. eng)

Rigg KK, Ford JA (2014) The misuse of benzodiazepines among adolescents: psychosocial risk factors in a national sample. Drug Alcohol Depend 137:137–42 (PubMed PMID: 24582910. Epub 2014/03/04. eng)

Robin C, Trieger N (2002) Paradoxical reactions to benzodiazepines in intravenous sedation: a report of 2 cases and review of the literature. Anesthesia Prog 49(4):128–32 (PubMed PMID: 12779114. Pubmed Central PMCID: PMC2007411. Epub 2003/06/05. eng)

Ross J, Darke S (2000) The nature of benzodiazepine dependence among heroin users in Sydney, Australia. Addiction 95(12):1785–1793

Ross J, Teesson M, Darke S, Lynskey M, Ali R, Ritter A, et al (2005) The characteristics of heroin users entering treatment: findings from the Australian treatment outcome study (ATOS). Drug and Alcohol Rev 24(5):411–418

Rudorfer MV, Osman OT, Potter WZ (1989) Alprazolam and aggression. Am J Psychiatry 146 (7):949–950

Schweizer E, Rickels K, Case WG, Greenblatt DJ (1991) Carbamazepine treatment in patients discontinuing long-term benzodiazepine therapy: effects on withdrawal severity and outcome. Arch Gen Psychiatry 48(5):448–452

Short- and Long-Term Use of Benzodiazepines in Patients with Generalized Anxiety Disorder (2014) A review of guidelines. Canadian Agency for Drugs and Technologies in Health, Ottawa ON

Simon GE, Ludman EJ (2006) Outcome of new benzodiazepine prescriptions to older adults in primary care. Gen Hosp Psychiatry 28(5):374–378 (PubMed PMID: 16950371. Pubmed Central PMCID: PMC2262838. Epub 2006/09/05. eng)

Soumerai S, Simoni-Wastila L, Singer C, Mah C, Gao X, Salzman C, et al (2003) Lack of relationship between long-term use of benzodiazepines and escalation to high dosages. Psychiatr Serv, 1006–1011

Stitzer ML, Griffiths RR, McLellan AT, Grabowski J, Hawthorne JW (1981) Diazepam use among methadone maintenance patients: patterns and dosages. Drug Alcohol Depend 8(3):189–199

Stone AM, Merlo LJ (2011) Attitudes of college students toward mental illness stigma and the misuse of psychiatric medications. J Clin Psychiatry 72(2):134–139 (PubMed PMID: 21208582. Pubmed Central PMCID: PMC3056282. Epub 2011/01/07. eng)

Substance Abuse and Mental Health Services Administration (2011) Treatment Episode Data Set (TEDS). 1999–2009. National Admissions to Substance Abuse Treatment Services, DASIS Series: S-56, HHS Publication No. (SMA) 11-4646, Rockville, MD; Substance Abuse and Mental Health Services Administration, 2011

Substance Abuse and Mental Health Services Administration (2014) Benzodiazepines in combination with opioid pain relievers or alcohol: greater risk of more serious ED visit outcomes, 18 Dec 2014

Therapeutic Guidelines Limited (2015) TG complete: psychotropic guidelines. Melbourne. Available from: http://etg.tg.com.au/complete/

Tone A (2005) Listening to the past: history, psychiatry, and anxiety. Can J Psychiatry 50(7):373–80 (PubMed PMID: 16086534. Epub 2005/08/10. eng)

Tsimtsiou Z, Ashworth M, Jones R (2009) Variations in anxiolytic and hypnotic prescribing by GPs: a cross-sectional analysis using data from the UK quality and outcomes framework. Br J Gen Pract J Roy Coll Gen Pract 59(563):e191–8 (PubMed PMID: 19520017. Pubmed Central PMCID: 2688068)

Verster JC, Volkerts ER (2004) Clinical pharmacology, clinical efficacy, and behavioral toxicity of alprazolam: a review of the literature. CNS Drug Rev 10(1):45–76

Verster JC, Volkerts ER, Verbaten MN (2002) Effects of alprazolam on driving ability, memory functioning and psychomotor performance: a randomized, placebo-controlled study. Neuropsychopharmacol 27(2):260–9

Victorri-Vigneau C, Gerardin M, Rousselet M, Guerlais M, Grall-Bronnec M, Jolliet P (2014) An update on zolpidem abuse and dependence. J Addict Dis 33(1):15–23 (PubMed PMID: 24467433. Epub 2014/01/29. eng)

Voyer P, Preville M, Cohen D, Berbiche D, Beland SG (2010) The prevalence of benzodiazepine dependence among community-dwelling older adult users in Quebec according to typical and atypical criteria. Can J Aging = La revue canadienne du vieillissement 29(2):205–213 (PubMed PMID: 20420748. Epub 2010/04/28. eng)

Walker BM, Ettenberg A (2001) Benzodiazepine modulation of opiate reward. Exp Clin Psychopharmacol 9(2):191–197 (PubMed PMID: 11518095. Epub 2001/08/24. Eng)

Wick JY (2013) The history of benzodiazepines: the Consultant pharmacist. J Am Soc Consultant Pharmacists 28(9):538–548 (PubMed PMID: 24007886. Epub 2013/09/07. eng)

Woods JH, Katz JL, Winger G (1992) Benzodiazepines: use, abuse, and consequences. Pharmacol Rev 44(2):151–347 (PubMed PMID: 1356276. Epub 1992/06/01. eng)

Yen CF, Ko CH, Chang YP, Yu CY, Huang MF, Yeh YC, et al (2014) Dependence, misuse, and beliefs regarding use of hypnotics by elderly psychiatric patients taking zolpidem, estazolam, or flunitrazepam. Asia-Pac Psychiatry Official J Pac Rim Coll Psychiatrists (PubMed PMID: 25296384. Epub 2014/10/09. eng)

Zador D, Sunjic S (2000) Deaths in methadone maintenance treatment in New South Wales, Australia 1990–1995. Addiction 95(1):77–84

Recent Trends in Alcohol and Other Drug Use Among Police Detainees in New Zealand, 2010–2015

Chris Wilkins, Jitesh Prasad, Karl Parker, Marta Rychert and Helen Moewaka Barnes

Abstract *Background*: New Zealand has unusual patterns of recreational substance use by international standards including low levels of cocaine and heroin use, and high methamphetamine use. *Aims*: This paper examines recent trends in alcohol and other drug use among police detainees in New Zealand over the past six years. *Method*: The paper utilises data from the New Zealand Arrestee Drug Use Monitoring (NZ-ADUM) study. NZ-ADUM interviewed approximately 800 police detainees each year at four central city police watch houses (i.e. Whangarei, Auckland, Wellington, Christchurch) from 2010 to 2015. *Results*: The proportion of police detainees who had used methamphetamine in the previous year increased from 28% in 2012 to 36% in 2015. Drinking prior to arrest declined from 41% in 2013 to 28% in 2015. The use of cannabis in the past year declined slightly from 76% in 2011 to 69% in 2015. The proportion using ecstasy in the previous year steadily declined from 28% in 2011 to 19% in 2015. Only small minorities had recently used cocaine or an opioid. Use of methamphetamine and ecstasy increased in Christchurch. *Conclusion*: Growing methamphetamine use is consistent with record seizures of methamphetamine over the past 2–3 years. Increasing drug use in Christchurch may reflect factors related to the devastating earthquakes in 2011 and the subsequent city rebuild, including an influx of construction workers, more organised trafficking groups and earthquake-related stress. The decline in cannabis use may be related to the emergence of 'legal' synthetic cannabinoids. The decline in ecstasy use may be the result of recent domestic enforcement operations and the overall global shortage of MDMA. The decline in alcohol drinking may be due to the introduction of pre-charge formal warnings for minor alcohol and disorder offences, and new restrictions on alcohol premise opening hours.
Acknowledgements: The New Zealand Drug Use Monitoring (NZ-ADUM) research study is funded by the New Zealand Police and is conducted by SHORE and Whariki Research Centre, College of Health at Massey University, Auckland. We would like to thank New Zealand Police staff at Whangarei, Auckland Central,

C. Wilkins (✉) · J. Prasad · K. Parker · M. Rychert · H.M. Barnes
SHORE and Whariki Research Centre, College of Health, Massey University, Auckland, New Zealand
e-mail: C.Wilkins@massey.ac.nz

Curr Topics Behav Neurosci (2017) 34:161–172
DOI 10.1007/7854_2016_471
Published Online: 23 April 2017

Wellington Central and Christchurch Central police watch houses for their assistance and cooperation with this research. We would also like to thank all the interviewers who worked with us on NZ-ADUM and all the police detainees who agreed to be interviewed for the study. The views expressed in this paper are entirely our own and do not necessarily reflect those of New Zealand Police.

Keywords Methamphetamine · Cannabis · Alcohol · Police detainees · New Zealand

Contents

1 Introduction

New Zealand has unusual patterns of recreational drug use compared to many other developed Western nations (Wilkins et al. 2002). The use of cocaine and heroin has historically been very low, while methamphetamine has been one of the most widely used illegal drugs since the early 2000s (Wilkins et al. 2015a, b). The ecstasy available in New Zealand is expensive by international standards and of uncertain composition, often containing a range of new psychoactive substances rather than MDMA. This unusual drug use environment is attributed to specific geographical and demographic characteristics of New Zealand including its geographic isolation from main international cocaine and heroin trafficking routes, its island geography, small population, and history of tight border controls to protect the primary agricultural sector (New Zealand Customs Service 2002). As a consequence, drug types which can be produced locally, such as cannabis and methamphetamine, are more readily available than those which must be smuggled internationally from distant producer countries, such as cocaine and heroin (Wilkins et al. 2015a, b).

Police detainees are a sentinel group for monitoring trends in substance use as they have high levels of alcohol and other drug use, high contact with criminal networks and illegal drug markets, and high levels of substance-related problems, such as problem drinking and drug dependency (Gaffney et al. 2010). They are also a vulnerable population with high incidences of mental health problems and developmental disadvantage, including parental neglect and youth delinquency (Wilkins et al. 2012). This paper examines recent trends in alcohol and other drug use among police detainees in New Zealand over the past six years.

2 Method

The paper utilises data from the New Zealand Arrestee Drug Use Monitoring (NZ-ADUM) study. NZ-ADUM has interviewed approximately 800 police detainees each year at four central city police watch houses (i.e. Whangarei, Auckland Central, Wellington Central and Christchurch Central) from March to August since 2010 (i.e. 809 = 2010, 828 = 2011, 800 = 2012, 848 = 2013, 832 = 2014, 835 = 2015). NZ-ADUM interviewers were present at these four police watch houses during the morning and evening shifts on every day of the week for the five-month study period each year. The interviewing times were selected to match the two periods of the day when the cells were at their fullest (i.e. following the night shift and following the day shift). It is not ethical, safe, or practical to interview some detainees due to their violent behaviour, intoxication, or emotional state. Consequently, detainees were excluded from the study if they were:

- under 17 years of age;
- unfit for interview due to intoxication from alcohol/drugs or medications;
- unfit for interview due to mental health issues;
- unable to understand the questions due to poor English language comprehension;
- unfit for interview due to threatening or violent behaviour;
- held in custody for more than 48 h;
- deemed unavailable by watch house staff due to ongoing legal/administrative proceedings.

Those detainees who were eligible to participate in the study were escorted to a private interview room where the NZ-ADUM interviewer introduced themselves as an independent university researcher, explained the aims of the study, and invited the detainee to participate. The interviewer explained that participation in the study was voluntary, everything they said would be confidential, no individual information would be shared with police, the results of the study would only be reported in aggregate, and they could choose not to answer any question if they did not want to. The ethical protocols used in NZ-ADUM have been approved by the Massey University Human Subjects Ethics Committee.

3 Analysis

The 2011, 2012, 2013, 2014, and 2015 NZ-ADUM survey waves were weighted to match the locational distribution of interviews completed in 2010 to ensure consistent comparisons over time. The number of interviews completed in each site location has generally been fairly similar from year to year, so the impact of the weighting is low. Statistical comparisons were made over the six years (i.e. 2010–2015), and between the four regional sites of the study. When a statistically significant difference was found over the six years, additional tests were conducted to compare specific years to each other, with the *p*-values adjusted for multiple comparisons using the simulation method in SAS. Differences between proportions (e.g. used cannabis in previous year) were tested using logistic regression. Some continuous variables were highly positively skewed (e.g. frequency of drug use and number of alcoholic drinks consumed); hence, statistical testing was run on the log-transformed values for these items to reduce the influence of outliers. All analyses were run using SAS version 9.3.

4 Demographics

The police detainees were overwhelmingly male, younger, unemployed or on a sickness benefit, and had fairly low levels of educational achievement (Table 1). Approximately 40% of the detainee sample was Maori compared to 15% of the general New Zealand population. The proportion of detainees who had completed the compulsory years of high school education increased from 47% in 2010 to 64% in 2015 ($p < 0.0001$). There were no other statistically significant changes in the demographic characteristics of the sample from 2010 to 2015. In 2015, 29% of the

Table 1 Selected demographic characteristics of police detainees, 2010–2015

Demographic	2010	2011	2012	2013	2014	2015
Male (%)	89	87	86	85	86	86
Age (years)	28	28	28	29	29	28
Maori (%)	38	40	40	41	43	42
European (%)	44	39	45	41	40	40
Unemployed/sickness benefit (%)	56	55	55	54	52	51
Completed high school (%)	47	54	58	54	57	63***
Ever mental illness (%)	32	30	34	35	36	29
Currently receiving treatment/medication for mental illness (%)	12	9	10	9	8	8

***$p < 0.0001$

detainees had suffered from a mental illness at some stage in their lives, and 8% were currently receiving treatment or medication for a mental illness at the time of their arrest.

5 Methamphetamine Use

The proportion of detainees who had used methamphetamine in the previous year increased from 28% in 2012 to 36% in 2015 ($p = 0.0047$). The mean number of days users had used methamphetamine increased from 68 days in 2010 to 89 days in 2015 ($p < 0.0001$). Important differences in methamphetamine trends were found between the study locations. The proportion of detainees in Christchurch Central who had used methamphetamine in the past year increased steadily from 20% in 2012 to 33% in 2015 ($p = 0.0036$) (Fig. 1). The mean number of days methamphetamine users in Christchurch had used methamphetamine increased from 58 days in 2012 to 94 days in 2015 (p < 0.0001). The proportion of detainees in Wellington Central who had used methamphetamine increased sharply from 28% in 2014 to 43% in 2015 ($p = 0.0028$).

In contrast, methamphetamine use remained largely stable in Auckland. The proportion of detainees from Auckland Central who had used methamphetamine in the previous 12 months remained unchanged from 38% in 2011 to 35% in 2015.

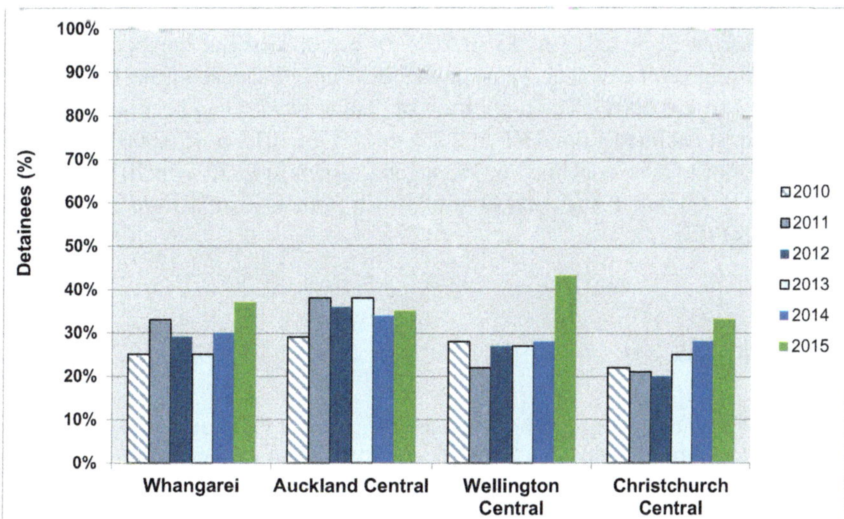

Fig. 1 Proportion of police detainees who used methamphetamine in the past 12 months by location, 2010–2015

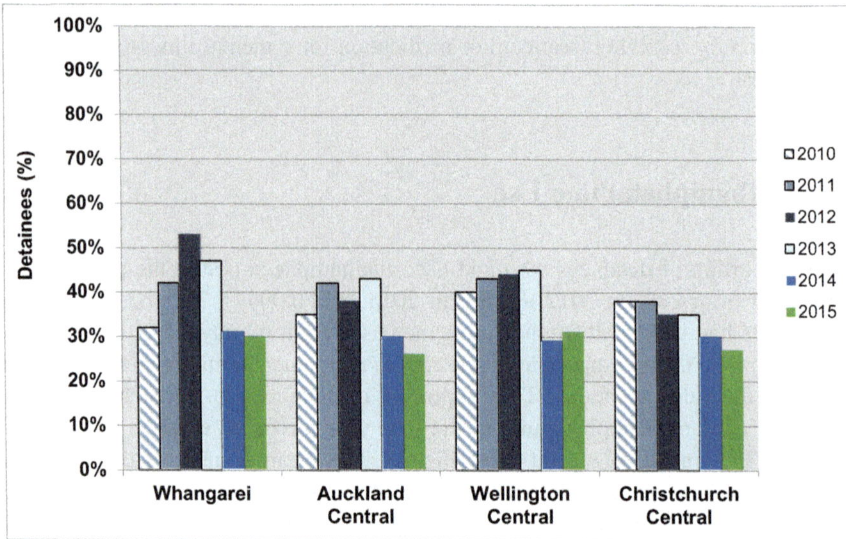

Fig. 2 Proportion of police detainees who had been drinking alcohol prior to their arrest by location, 2010–2015

6 Alcohol Consumption

The mean number of days on which the police detainees had drunk alcohol in the previous year declined from 101 days in 2013 to 82 days in 2015 ($p = 0.0070$). The mean number of standard alcoholic drinks consumed on a typical day of use decreased from 18 in 2013 to 15 in 2015 ($p = 0.0003$). The number of standard drinks consumed on a typical occasion had previously increased from 12 in 2010 to 18 in 2013 ($p < 0.0001$). The proportion of detainees who had been drinking prior to their arrest declined from 41% in 2013 to 28% in 2015 ($p < 0.0001$). Levels of drinking prior to arrest declined in Whangarei (down from 53% in 2012 to 30% in 2015, $p < 0.0001$) and Auckland Central (down from 43% in 2013 to 26% in 2015, $p = 0.0024$) (Fig. 2).

7 Cannabis Use

The proportion of detainees who had used cannabis in the previous year decreased from 76% in 2011 to 69% in 2015 ($p = 0.0219$). The mean number of days the detainees had used cannabis in the previous year declined from 187 days in 2010 to 155 days in 2015 ($p = 0.0052$). The reduction in cannabis use was strongest in

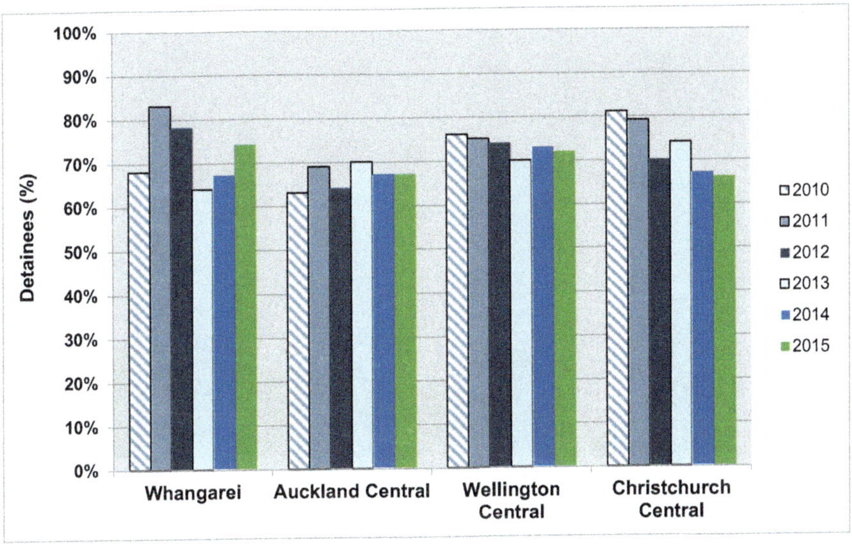

Fig. 3 Proportion of police detainees who had used cannabis in the past 12 months by location, 2010–2015

Christchurch. The proportion of detainees in Christchurch Central who had used cannabis in the past year declined from 79% in 2011 to 66% in 2015 ($p = 0.0230$) (Fig. 3). There appeared to be some rebound in cannabis use among the detainees in 2015. The proportion of detainees in Whangarei who had used cannabis in the previous month increased from 45% in 2013 to 63% in 2015 ($p = 0.0227$).

8 Ecstasy Use

The proportion of detainees who had used ecstasy in the previous year declined steadily from 28% in 2011 to 19% in 2015 (p = 0.0003). The proportion of Whangarei detainees who had used ecstasy in the previous year fell substantially from 36% in 2011 to 6% in 2015 ($p < 0.0001$) (Fig. 4).

In contrast to the overall picture of declining ecstasy use, the proportion of Christchurch Central detainees who had used ecstasy in the past year increased from 14% in 2014 to 24% in 2015 ($p = 0.0169$). The number of days detainees in Christchurch Central had used ecstasy in the previous year increased from five days in 2010 to 14 days in 2015 ($p = 0.0091$).

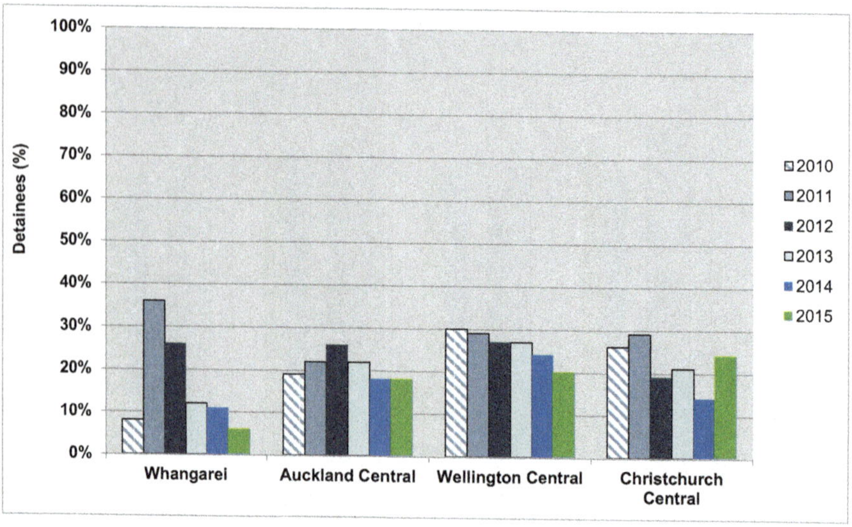

Fig. 4 Proportion of police detainees who had used ecstasy in the past 12 months by location, 2010–2015

9 Cocaine Use

The proportion of detainees who had tried cocaine at some point in their lives has been steadily increasing in recent years, up from 17% in 2010 to 24% in 2015 ($p = 0.0091$), but the proportion who report using it in the previous year has remained persistently low and largely unchanged (i.e. 4% = 2010, 4% = 2011, 5% = 2012, 5% = 2013, 6% = 2014, 5% in 2015).

10 Opioid Use

There was no change in the proportion of detainees who had used an opioid in the previous 12 months from 2010 to 2015 (i.e. 8 to 6%). In 2015, detainees in Christchurch Central were more likely to have used an opioid in the previous 12 months than detainees in Whangarei (11% vs. 2%, $p = 0.0094$) and Auckland Central (11% vs. 4%, $p = 0.0113$). The higher prevalence of opioid use in Christchurch compared to the other study locations has been evident in every year of the study.

11 Discussion

The police detainees reported high levels of drinking and other drug use compared to the wider population, confirming the value of studying this population as a means to monitor wider trends in substance use and related harm. For example, 36% of the police detainees reported using methamphetamine in the previous year compared to 0.9% of the general New Zealand population (aged 16–64 years) (Ministry of Health 2015). The increase in methamphetamine use by the detainees in recent years is consistent with the record seizures of methamphetamine made by the New Zealand Police and New Zealand Customs Service over the past three years (i.e. 31 kg in 2013; 99 kg in 2014; 334 kg in 2015) (NDIB 2016). In June 2016, New Zealand Police and Customs made a record one-off 448-kg seizure of methamphetamine of (34% more than the weight of methamphetamine seized in the entire previous year). Record seizures of methamphetamine have also been made at the Australian border in recent years (ACC 2015), and there have been growing seizures of methamphetamine in Europe (EMCDDA 2016). The United Nations Office of Drug Control (UNODC) report that global seizures of methamphetamine doubled from 2009 to 2013, and in their most recent World Drug Report contend that methamphetamine supply is increasingly globally interconnected (UNODC 2015).

The NZ-ADUM findings also highlight the importance of regional differences in drug trends, and this local variation can be overlooked when the focus is so often on national drug prevalence measures. The use of methamphetamine among Christchurch detainees has increased steadily since 2012, and this is broadly consistent with findings from other drug monitoring sources. The Illicit Drug Monitoring System (IDMS), which interviews 300 frequent illegal drug users each year in Auckland, Wellington, and Christchurch, reported the availability of methamphetamine recovered sharply in Christchurch in 2013, following a number of years of decline, and this resurgence continued in 2014 (Wilkins et al. 2015b). One possible explanation for the surge in methamphetamine use in Christchurch is that it has been driven by the influx of construction workers to Christchurch over recent years for the city rebuild, following the devastating earthquakes there in 2010 and 2011. The reconstruction effort in Christchurch city involves thousands of construction and related workers, mostly young men, who will be earning good money and, as a population group, have higher levels of amphetamine use (see Ministry of Health 2015). New Zealand Police have also reported that local Christchurch criminal gangs have recently been absorbed into larger national gangs with a greater focus on methamphetamine trafficking (NDIB 2015), possibly in response to the rise in demand for drugs in Christchurch. The IDMS found the proportion of frequent drug users who purchased methamphetamine from a gang member increased from 36% in 2013 to 50% in 2014 (Wilkins et al. 2015b). Earthquake-related stress may also be a factor in rising drug use in Christchurch. In contrast, methamphetamine use among the detainees in Auckland Central has been stable over recent years. This is important to note as Auckland has traditionally been the largest methamphetamine market in New Zealand.

The reduction in alcohol drinking among the detainees was surprising following a number of years of increasing alcohol consumption but may reflect a number of policy and enforcement initiatives including the greater use of Pre-charge Warnings (PCW) for minor alcohol offences, and the impact of new government restrictions on alcohol premise opening hours. PCW allow offenders to be arrested and taken to a police station but instead of proceeding with an expensive and time consuming prosecution the offender is formally warned and released (O'Reilly 2010). New Zealand Police have noted over half of the PCW issued were for disorderly behaviour and breaches of liquor bans (New Zealand Police 2013). The greater use of PCW for minor alcohol offences may have meant fewer heavy drinkers were detained in the cells overnight and hence available to be interviewed for NZ-ADUM.

One explanation for the decline in the use of cannabis among the detainees is the greater availability of synthetic cannabinoids in New Zealand in recent years since 2010. Synthetic cannabinoid products are often marketed as 'legal' alternatives to cannabis and are not detectable in standard drug testing assays (Perrone et al. 2013). The latter attribute may make synthetic cannabinoids particularly attractive to police detainees who, as a group, are more likely to be subject to drug testing as part of parole and home detention conditions. In a recent analysis of drug substitution related to legal high use among police detainees, 94% of synthetic cannabinoid using detainees who said they had substituted a drug reported they had substituted synthetic cannabinoids for natural cannabis (Wilkins et al. 2016). Consistent with this explanation, there is some evidence of a return to cannabis use by detainees following the ban of all synthetic cannabinoids in 2014. Another possible explanation is the increasing effectiveness of the cannabis crop eradication operations. New Zealand Police have indicated there is now a greater focus on organised criminal groups involved in cannabis cultivation as part of these operations, and this may have negatively impacted cannabis supply (NDIB 2016).

The reduction in ecstasy use may reflect the impact of local enforcement operations against domestic New Zealand 'ecstasy' syndicates in 2011/12, and the overall global shortage of MDMA since around 2009 (EMCDDA 2016). The global shortage of MDMA created a market gap for domestic New Zealand criminal syndicates to start manufacturing fake 'ecstasy', containing blends of cathinones and piperazines (ESR 2014) on a large scale, and this significantly increased the availability of 'ecstasy' to many regional areas in New Zealand where ecstasy was usually rare. A series of enforcement operations dismantled these syndicates in 2011/12 (NDIB 2013). The recent increase in ecstasy use by detainees in Christchurch is consistent with the impact of the influx of construction workers on the city rebuild, some of whom will be from countries where ecstasy is more widely available.

The increasing lifetime experience of cocaine use with little change in current use may reflect the fact that cocaine used by detainees is largely obtained outside of New Zealand, for example during overseas holidays and work. The higher level of opioid use in Christchurch compared to the other study locations is consistent with the long-standing situation of a large injecting drug using population in Christchurch (Newbold 2000).

There are important limitations to consider when assessing the findings from NZ-ADUM in regard to assessing wider alcohol and drug trends. As outlined early, some detainees cannot be safely interviewed due to intoxication, violent behaviour, and emotional state, and these particularly problematic detainees may have different substance use patterns. Secondly, the detainee sample has distinct demographic and socioeconomic characteristics, and this may influence the type of drug types they have access to and are likely to use. For example, it has been suggested that the high price of cocaine in New Zealand is likely to limit its use to higher socioeconomic groups. Thirdly, police detainees may be particularly attracted to new psychoactive products, such as synthetic cannabinoids, because they cannot be detected in the drug testing required in the criminal justice system. Finally, it could be argued that detainees may be reluctant to self-report drug use while in custody. However, the detainees did self-report high levels of illegal drug use, including drug types such as methamphetamine which attract high criminal penalties and stigma, so while underreporting is likely it did not prevent honest reporting in many instances. Furthermore, our analysis involved comparisons of samples of detainees between years. There is little reason to believe the incentive to under-report use would have dramatically changed over time or apply to one drug, for example in the case of declining cannabis use, but not another, as in the case of increasing methamphetamine use.

References

ACC (2015) Australian Crime Commission's illicit drug data report 2013–14. Australian Crime Commission, Canberra

EMCDDA (2016) European drug report 2016: trends and developments [European Monitoring Centre for Drugs and Drug Addiction]. Available at http://www.emcdda.europa.eu/system/files/publications/2637/TDAT16001ENN.pdf. Accessed 16 June 2016

ESR (2014) ESR drugs trends report February 2013–October 2013. Institute of Environmental Science and Research, Wellington

Gaffney A, Jones W, Sweeney J, Payne J (2010) Drug use monitoring in Australia: 2008 annual report on drug use among police detainees. Monitoring Report No. 9. Australian Institute of Criminology

Ministry of Health (2015) Amphetamine use 2014/15. Wellington, New Zealand Health Survey

NDIB (2013) Personal communication. National Drug Intelligence Bureau, Wellington

NDIB (2015) Personal communication. National Drug Intelligence Bureau, Wellington

NDIB (2016) Personal communication. National Drug Intelligence Bureau, Wellington

New Zealand Customs Service (2002) Review of customs drug enforcement strategies 2002. Project Horizon Outcome Report, Wellington

New Zealand Police (2013) Police adult diversion scheme. Wellington

Newbold G (2000) Crime in New Zealand. Dunmore Press, Palmerston North

O'Reilly J (2010) New Zealand police pre-charge warnings alternative resolutions evaluation report. New Zealand Police, Wellington

Perrone D, Helgesen R, Fischer R (2013) United States drug prohibition and legal highs: how drug testing may lead cannabis users to Spice. Drugs Edu Prev Policy 20:216–224
UNODC (2015) World drug report 2015. United Nations Office on Drugs and Crime, Vienna
Wilkins C, Casswell S, Bhatta K, Pledger M (2002) Drug use in New Zealand: national surveys comparison 1998 and 2001. Alcohol and Public Health Research Unit, Auckland
Wilkins C, Parker K, Prasad J, Jawalkar S (2016) Do police arrestees substitute legal highs for other drugs? Int J Drug Policy (in press 18 January)
Wilkins C, Prasad J, Parker K, Moewaka Barnes H, Asiasiga L, Rychert M (2015a) New Zealand Arrestee Drug Use Monitoring (NZ-ADUM) 2010–2014. SHORE and Whariki Research Centre, College of Health, Massey University, Auckland
Wilkins C, Prasad J, Wong K, Rychert M (2015b) Recent trends in illegal drug use in New Zealand 2006–2014: findings from the 2006, 2007, 2008, 2009, 2010, 2011, 2012, 2013 and 2014 illegal drug monitoring system (IDMS). SHORE and Whariki Research Centre, Massey University, Auckland
Wilkins C, Sweetsur P, Moewaka Barnes H, Smart B, Asiasiga L, Warne C (2012) New Zealand Arrestee Drug Use Monitoring (NZ-ADUM)—2011 results. SHORE and Whariki Research Centre, School of Public Health, Massey University, Auckland

The Impact of Legalizing and Regulating Weed: Issues with Study Design and Emerging Findings in the USA

Priscillia E. Hunt and Jeremy Miles

Abstract Evaluations of the impact of medical and recreational marijuana laws rely on quasi- or natural experiments in which researchers exploit changes in the law and attempt to determine the impact of these changes on outcomes. This chapter reviews three key issues of causal inference in observational studies with respect to estimating of impact of medical or recreational laws on marijuana use—intervention definition, outcome measurement, and random assignment of study participants. We show that studies tend to use the same statistical approach (differences-in-differences) and yet find differential impacts of medical marijuana laws on adult use in particular. We demonstrate that these seemingly conflicting findings may be due to different years of analysis, ages of the study sample in each year, and assignment of jurisdictions to the control group versus treatment group.

Keywords Marijuana legalization · Medical marijuana laws · Cannabis use · Causal inference

Contents

P.E. Hunt (✉) · J. Miles
RAND Corporation, 1776 Main Street, Santa Monica, CA 90407, USA
e-mail: phunt@rand.org

P.E. Hunt
Institute for the Study of Labor (IZA), Bonn, Germany

Curr Topics Behav Neurosci (2017) 34:173–198
DOI 10.1007/7854_2015_423
Published Online: 31 December 2015

1 Introduction

Changes in marijuana policies for medical and recreational purposes in the USA present an exciting research opportunity. The policy variation over time and across states is substantial, which means researchers may be able to detect the impact of marijuana policies on the use of, and harms associated with, marijuana, if they exist.[1] While the prospect of empirical findings is enticing, there is a need to take a step back to understand what we are analyzing and what it all means.

The word 'cause' is thrown about carelessly in everyday conversation. In the overlapping world of philosophy and statistics, the notion of cause is considerably more complex (Russell 1912; Kenny 1979; Holland 1986; Pearl 2000). At the risk of offending these scholars, we shall skip over the complexities of the definition of 'cause' and say that if changes in policy give rise to changes in outcomes, then the outcomes were caused by the policy change. The gold standard for establishing causation is the randomized control trial (RCT)—in which individuals are randomly assigned to one of two (or more) conditions. Any differences in outcomes between participants in the two groups can be attributed to the effect of the treatment.[2] Randomized trials grew out of agricultural research in the early twentieth century, thanks to the work of Fisher (1954), and were adopted by fields of psychology and education (Guilford 1942) and then slightly later by medicine; the first randomized trial in health is generally considered to be Medical Research Council (1948).

The great advantage of agricultural research, where RCTs were developed, is that both measurement and randomization are straightforward—if one is researching whether changing the feed increases the growth rate of pigs, the pigs do not complain about the feed they are given, and it is clear what the outcome is (the weight of the pig) and how it is assessed (by weighing the pig before and after). In addition, the pig does not object to the process of measurement, as it is usually deceased. In psychology, education, and medicine, randomized trials are a little more difficult, but still relatively straightforward. Individuals are randomized to an intervention or control group, and outcomes are measured. There is greater probability of complications arising than in agricultural research because people may

[1]More specifically, all null hypotheses (at least in their two-tailed forms) are false (Cohen 1988); therefore, a change in policy must have an effect—the question is the size and direction of the effect.

[2]Although the RCT is considered the 'gold standard' for establishing causation, it is interesting to note that the first RCT in medicine was published as recently as 1948.

decline treatment or refuse to be measured, for example, but the process is feasible and clear.

In policy research, individuals clearly cannot be randomized to have different laws applying to them, and so our attempt at establishing causality through the use of randomized trials breaks down before it starts. Instead, we can rely on quasi- or natural experiments, where we take advantage of changes that are occurring, and attempt to determine the impact of these changes (Campbell and Stanley 1963). When developing research designs to identify the causal impact of a policy or program, there are three ingredients that differ between the gold standard randomized trial and the natural experiment that we will consider: defining the intervention, measurement of outcomes, and random assignment of study participants.

First, we define the intervention of interest that determines the difference between the intervention (or treatment) group and the control group. The treatment group receives the intervention, and the control group does not (e.g., treatment as usual, or usual care). In the case of a randomized trial, we can formulate the intervention and operationalize it precisely. For example, our intervention might be the addition of one ton of pig manure per acre of cabbage field. In the area of policy research, this is not possible—we might want to know the effect of, for example, removal of criminal justice restrictions on supply and possession of marijuana for recreational purposes. We cannot implement this change in a precise manner—we must take what is offered to us in terms of changes in the law and attempt to use these changes in the law and establish their effects. Recent research has shown that there is a great deal of heterogeneity in the detail and implementation of laws that seem, on the surface, to be similar (Pacula et al. 2014) and that this heterogeneity may, or may not, be important for health and safety purposes (Pacula et al. 2015; Anderson and Rees 2014; Sevigny et al. 2014).

Second, we consider the measurement of outcomes. In randomized trials in agriculture, medicine, or psychology, the outcomes of interest can be defined prior to the study taking place. For example, we may want to assess whether using one type of pesticide leads to larger cabbages; we may want to know whether using one type of drug leads to more patients surviving; we may want to know whether using one type of teaching method leads to more students graduating. Regarding marijuana policy, this means the outcomes of interest should be defined and measured prior to the passing of a law, so that a comparison can be made afterward. For example, we may want to know whether passing a MML leads to more traffic accidents, marijuana use, or alcohol use, so we need to measure the number of accidents and the proportion of people using marijuana and/or alcohol. While this may seem obvious, there can be issues with an intervention affecting the type of outcome measured. That is, we know that changes in the law will have effects, but we do not always know in advance what those effects will be. Marijuana legalization and regulation could lead to product innovation and changes in consumption behavior, as well as other unexpected consequences. In which case, studies could show consumption going down if some people switch or start consuming edibles or vaping and we only assessed more conventional smoking. This is not necessarily a problem, but it does mean the researcher needs to be clear about what is and is not being measured in the study and what conclusions can be drawn.

Finally, we consider randomization. Clearly, we cannot randomize individuals to have different laws apply to them—and we cannot randomize states or other jurisdictions (e.g., change borders). However, consider the purpose of randomization—it is to ensure that, at the population level, there are no differences between the group who receive the intervention and the group who do not. Jurisdictions are not randomly assigned a law—individuals or members of legislatures vote for laws given their preferences, which are likely to be correlated with outcomes of interest (e.g., levels of marijuana use). Since voting for marijuana is a big step, these states likely differ in important ways from states that vote against marijuana and states that never have the opportunity to vote on marijuana (because there is not enough support to bring the item on the ballot). However, when researchers measure the outcome variable prior to the passage of the law, it is possible to use statistical methods that adjust for these differences and provide estimates of the effect of the intervention adjusting for baseline differences.

The most common approaches—difference-in-differences (DiDs) (Ashenfelter and Card 1985), regression discontinuity (RD) (Lee and Lemieux 2010), propensity score matching (Rosenbaum and Rubin 1983), instrumental variable (Angrist et al. 1996), and synthetic control approach (Abadie et al. 2010)—aim to remove the selectivity bias of non-random assignment. In addition to describing studies comprising the growing body of literature using these approaches to estimate the impact of MMLs, we describe why there are currently no evaluations of recreational laws. This paper starts by briefly describing the modern history of medical marijuana laws (MMLs) and recreational marijuana laws (RMLs) in the USA. We then describe issues with study design in identifying the impact of these marijuana laws and provide initial findings in the literature for both MMLs and RMLs in the USA. By providing empirical findings along with design issues, it is hoped that this paper helps tie together what the emerging empirical findings are telling us, and not telling us, and where we may be going in the future.

2 Background on the Modern Transition of US Medical and Recreational Laws

While the US federal government has prohibited marijuana since the 1937 Marijuana Tax Act, the modern history of MMLs goes back to 1970 when the US federal government passed the Control Substances Act (CSA). The CSA placed all controlled substances into five categories, or ‘Schedules,’ based on the relative potential for abuse as well as recognized medical usefulness. Marijuana was classified as a Schedule I controlled substance, implying there were no currently accepted medical uses in the USA and making it a criminal offense for doctors to medically prescribe marijuana or cannabis and for individuals to distribute or possess marijuana.

Although there has been, and remains to be, issues between the federal and the state law, US states continue to pass laws increasing legal access to marijuana. While there are a number of differences in the actual provisions adopted (Pacula et al. 2014), the provisions of MMLs can be clustered in those focused on consumers, those focused on suppliers, and then opening access for recreational trade.

2.1 Protection of the Rights of Patients: 1996–2002

The modern transition of MMLs started in 1996 when California residents voted in favor of Proposition 215 (the Compassionate Use Act of 1996), which allowed for the cultivation, distribution, possession, and consumption of marijuana for medical purposes without the threat of state prosecution.[3] Washington State followed suit in 1998, followed by Alaska and Maine in 1999, and by October 2013, 21 states and the District of Columbia had legally effective laws on medical marijuana in place (NCSL 2013). In these early years, states focused on ensuring the access and availability of marijuana to patients. This meant including in the language of the law the type of legal protections conferred to various market actors (e.g., physician, caregiver, patient), the conditions for which marijuana could be recommended, and whether patients or caregivers could supply the marijuana.

A key provision in state law that allowed for the transition to allowing the possession of marijuana for medical purposes was the explicit inclusion of either legal privilege (protection against both trial and conviction) or affirmative defense (protection against conviction). States did not always coordinate these protections for various actors—physicians, patients, dispensaries, and caregivers—so there were some legal gray areas regarding whether people arrested would have protection from prosecution. Generally speaking, however, most states offered affirmative defense for patients and caregivers, and legal privilege for physicians; it must be emphasized that this has not always been the case, nor is it the case for all jurisdictions (Pacula et al. 2014).

While laws in the early years were clear about home cultivation, they were silent or ambiguous about the legality of obtaining marijuana from third-party vendors, such as dispensaries or collective arrangements. The silence in the laws made the lines between legal medical markets and illegal recreational markets blurry, and the ambiguity led to significant confusion and attention by law enforcement.

The laws also include language about the conditions for which marijuana could be recommended. Most medical marijuana states allow for cancer, glaucoma, and HIV/AIDS, but state laws tend to differ on the circumstances in which a doctor can recommend marijuana for pain (Pacula et al. 2014). Jurisdictions may permit the

[3]Arizona voters also passed a set of laws permitting the use of marijuana for medicinal purposes in 1996; however, the statute included language allowing physicians to prescribe MM (Pacula et al. 2014).

use of medical marijuana for pain, although the language defining the permissible uses of medical marijuana for pain may vary and can be categorized into one of the following: (1) pain caused by a diagnosable medical condition, (2) any pain of unknown causes, and (3) upon recommendation of a physician. Most jurisdictions only permit marijuana use for pain caused by a diagnosable medical condition (72 %, 13 of 18), while WA and California have more relaxed definitions allowing for any pain of unknown causes or a physician's recommendation, and Maryland, Montana, and D.C. have different structures in place affecting use for pain (Pacula et al. 2014).

2.2 Protection of Access to Supply for Patients: 2003–2012

Most medical marijuana states have always allowed for individuals to supply their own marijuana through home cultivation. However, it is not necessarily straightforward for everyone to grow marijuana in their home, and in 2003, California legislature passed a second law (CA Senate Bill 420) that more clearly provided legal protection to marijuana 'cooperatives' and dispensaries to grow and sell marijuana to qualified patients. Legal protections for wholesale growers and processors remained unclear and were left to the courts to handle. Many other medical marijuana states followed suit by explicitly indicating dispensaries were either permitted, or not permitted. By January 2012, according to the functional definitions of state statutes and constitutional amendments, approximately 70 % (13 out of 18) of jurisdictions permit home cultivation and just over half of jurisdictions allow for dispensaries (Pacula et al. 2014).

2.3 Opening Access for Recreational Purposes: 2012–????

On November 2012, voters in Colorado (CO) and Washington (WA) passed Amendments or Initiatives to legalize the supply and consumption of marijuana for recreational purposes for adults 21 years of age and older. Two years later on November 2014, voters of Oregon (OR) and Alaska (AK) followed suit by passing laws legalizing a recreational marijuana market, while voters of Washington D.C. also passed an Initiative in 2014, although there are many limitations since much of D.C. land is on federal land.

Currently, only CO and WA have operating markets for recreational purposes. Colorado opened the market to marijuana trade for recreational purposes in January 2014 and Washington opened in July 2014. Various stakeholders shaped legislations and regulations for the cultivation, manufacturing, distribution, and testing of marijuana, and the Department of Revenue in particular developed regulations on issuing new licenses, license fees, and tax collection processes (Room 2014; CODoR 2014; MPP 2013). In WA, the regulatory decision-making is conducted by the state Liquor Control Board, and the Board is supported by a consultancy group, BOTEC.[4] For a description of regulations in each state, see Room (2014).

3 Identifying the Causal Effect of Marijuana Policies

As described earlier, the gold standard approach to identifying the causal effect of a program or intervention is a RCT. In a RCT or experimental design, units are randomly assigned to a control group or treatment group and outcome data are collected both before and after the treatment. The reality is that there are important questions facing our society, such as the impact of legalizing marijuana for medical purposes on youth use, in which a randomized controlled trial is not feasible. In this section, we highlight these study design issues for estimating the impacts of MMLs and RMLs.

3.1 Issues in Defining the Intervention for Marijuana Policy Analysis

The first feature of research designs to estimate causal impacts is to use a control group and a treatment group. At first glance, it may seem that a control group refers to jurisdictions without marijuana laws and a treatment group is jurisdictions that have marijuana laws. However, the reality is more nuanced than this. As described earlier in the background section, states pass 'packages' of laws, and there is not one medical marijuana law or one recreational law, as all jurisdictions adopt slightly different legal provisions or regulatory standards. These differences are not a trivial, academic exercise with little practical relevance—studies are finding these individual features of the law have differential impacts on consumption (Pacula et al. 2014; Wen et al. 2015) and potency (Sevigny et al. 2014). Knowing the difference could help policymakers identify better 'policy packages' in terms of health and safety outcomes.

When it comes to evaluating a change in the law, it may not always be obvious what the 'treatment' is. Indeed, one study finds that the definition of decriminalization was not well understood and some jurisdictions were considered decriminalized when in fact they had not, which potentially led to erroneous conclusions about the impact of decriminalization (Pacula et al. 2005). Sometimes treatment refers to a jurisdiction in which a MML or RML was passed and made legally effective. Decision-makers in the criminal justice system, for example, may be interested to know whether MMLs or RMLs reduce marijuana-related arrests or improve case throughput. By removing the constraint of the law, the criminal

[4]For reports for WA, see http://liq.wa.gov/marijuana/botec_reports.

justice system could experience a change in how things operate, so the interest is to understand whether the change in the law itself impacts behaviors of individuals, law enforcement, prosecutors, etc. Sometimes treatment refers to observable physical changes on the ground as the result of passing a MML or RML. This definition of treatment aims to understand the mechanism of change, i.e., whether access to marijuana at dispensaries affects use. For MMLs in particular, Maryland (MD) is an interesting state in this respect. A law was passed in MD providing an affirmative defense for patients, physicians, and caregivers (Pacula et al. 2014); however, the law is very restrictive, so some authors conclude that MD does have a MML (Pacula et al. 2015; Lynne-Landsman et al. 2013) and other authors conclude that MD does not (Anderson et al. 2015; Wen et al. 2015; Harper et al. 2012). Since there are delays between when a law is passed, when it is made effective, and when it is implemented on the ground, researchers need to be careful to identify the relevant date of interest for the units in the treatment group.

One should also give careful consideration in identifying the control group. A fundamental problem of causal inference is that a unit of observation cannot be in both the intervention group and control group at the same time (Rubin 1974). While there are a number of statistical techniques to identify a relevant control group, which we will discuss later, there is also a conceptual, design issue. Take the case of analyzing RMLs, for example, where all the treatment states in the USA thus far have passed MMLs. What would it mean to use a control group that includes the universe of MML and non-MML states without a RML? Can the outcomes in non-MML jurisdictions where there is no legal supply infrastructure for marijuana, no culture for purchasing marijuana in stores, and no interactions between law enforcement and individuals about a legal form of marijuana represent what would have happened in jurisdictions that do have this place? It would seem that including non-MML jurisdictions could lead to over/underestimates of the true effect of what would have happened in the RML states if they had not legalized recreational marijuana.

3.2 Measurement Complications in Marijuana Policy Evaluation

Studies with a pre–post test design measure the change in an outcome measure associated with a changed situation or phenomenon. When analyzing the impact of the changing legal situation for medical and recreational marijuana possession and supply, a key seemingly simple research question is: What impact does legalization have on marijuana use? There can be many perspectives on what is meant by use and problems measuring relevant pre- and post-implementation data. One option is to study whether MMLs/RMLs have an effect on the proportion of people, young and/or adult, who say that they have ever used marijuana. One dataset typically used to analyze this question for both young people, aged 12–20, and adults, aged

21 years and older, is the National Survey on Drug Use and Health (NSDUH), which is a repeated cross-sectional, nationally and state representative survey considered the primary source of information on substance-use behavior for the US civilian, non-institutionalized population. The survey has been conducted annually since 2004, so one limitation is the availability of pre-implementation period data. One dataset typically used for evaluating youth use is the National Longitudinal Study of Youth (NLSY97), which is a nationally representative cohort sample of the US population (aged 12 through 16 in 1997). A key benefit of this survey data is that the same individuals are followed over time, and there are a variety of rich characteristics that can be used for matching approaches. However, these data have a number of limitations, well including that it is not state representative; the sample is constantly aging, so respondents of late adopting states are older than respondents of early adopting states, and there are no data for California before the passage of the first MML. Another survey of young people that can be used for this research question is the National Youth Risk Behavior Survey (YRBS), which has been conducted since 1997, uses a repeated cross-sectional design, and is nationally representative of 9th- through 12th-grade students in public and private schools throughout the USA. These survey data suffer from the limitation that not all states are covered in every survey year, and it is collected biannually. A third, national youth survey is Monitoring the Future (MTF), which has been surveying 8th-, 10th-, and 12th-grade students each year since 1991 (12th graders only since 1975) with an annual follow-up with a sample of each graduating class for a number of years after their initial participation. The MTF specifically asks about use in the previous 12 months, not 'ever.'

The research question regarding use may be more specifically whether MMLs/RMLs have an effect on the frequency of use (or chronic use or heavy use). The NSDUH, NLSY97, and YRBS include an item for the number of days in the previous 30 days the individual used marijuana. MTF asks respondents whether they use 'daily.'

Another type of use-related question may be to understand whether MMLs/RMLs alter the age of initiation. The NSDUH includes an item for using marijuana for the first time during the past year. While the NLSY97 also includes a question for age at first use, the limitation of the sample constantly aging can mean there are limited pre-period data when people are young enough to be trying marijuana for the first time. The YRBS asks respondents whether they tried marijuana before the age 13 years.

Yet another question may be whether MMLs/RMLs have an effect on the misuse (or dependency) of marijuana. The NSDUH has a set of questions for the DSM-IV diagnostic criteria. The Treatment Episode Data Set (TEDS) is collected annually by state from substance abuse agencies and contains almost all substance abuse treatment admissions that occur within the USA, excepting private facilities that treat non-publicly insured individuals and facilities not receiving any federal or state grant funding. Since the unit of observation is an admission, the same person may be readmitted more than once in a year. Information is collected for the primary, secondary, and tertiary substance of abuse reported at the time of admission, so researchers need to consider whether treatment has an effect on how marijuana might be recorded in this respect.

The question about whether legalization has an effect on use may specifically refer to the amount used in a setting or use-day (or heavy use). This is an area for which virtually no information is available (Kilmer et al. 2014). The datasets described thus far do not ask respondents how much marijuana is consumed during a use-day. There are items in NSDUH about the size and price of recent purchases; however, people share marijuana with others for free, so it is unlikely this is useful for understanding daily patterns of consumption intensity of individuals. There are grams per joint information available between 2000 and 2003 for the arrestee population in the Arrestee Drug Abuse Monitoring Program (Zhang 2003); however, not only are there problems with the years available, but also these data are for a select group of the population with less regard for the law. Not to mention, a researcher would need to make some assumptions about consumption amounts for people who do not smoke joints. Since it is unknown how legalization may affect the form of consumption (e.g., joints, bong, edible, vape), it is unlikely these data will provide reliable results for the particular research question.

There has been increased interest in understanding whether MMLs/RMLs have an effect on how marijuana is used (e.g., edibles, hash oil, joints). We are unaware of any dataset in the field that has consistently asked respondents over several years about how they consume marijuana. There are two datasets we are aware of with a relatively short time frame or very limited questions on other forms of use. The RAND West Coast Marijuana Use survey is a state representative, probability-based Web panel of non-institutionalized, 18+ year olds residing in four states (Washington, Oregon, Colorado, and New Mexico). There are currently three waves of data collected in October of 2013, May of 2014, and October of 2014. Respondents are asked whether the last time they used for medical purposes they consumed it by smoking, vaporizing, eating, drinking, or other; respondents are also asked the same question for recreational purposes. For synthetic cannabis in particular, the MTF asks respondents whether they have used in the last 12 months.

Last, the question may be whether MMLs or RMLs have effects on the strength of marijuana used (or potency in terms of tetrahydrocannabinol and cannabidiol). Similar to the amount consumed during a use-day, the national drug-use surveys (e.g., NSDUH, NLSY97) do not include items for the type, strain, or potency of marijuana. This may be because users are unlikely to provide reliable information. One dataset that is useful for other drug types is the System to Retrieve Information from Drug Evidence (STRIDE), which is a database of drug exhibits sent to Drug Enforcement Administration (DEA). However, marijuana observations are not chemically analyzed for potency (Kilmer et al. 2014). The Potency Monitoring Program (PMP) has performed forensic analysis of seized marijuana samples since 1970 and is a more reliable option for analyzing the impact of laws on marijuana potency [for more information on this dataset, see Sevigny (2013)].

Thus far, we describe measurement issues for data on marijuana use in both the pre-period and post-period, but the recent passage of RMLs has raised concerns about defining the post-period. At the time of this study, RMLs have only been in place in CO for 1.5 years and 1 year in WA, so any analyses using these data can only be identifying short-term effects that may not be permanent. This raises another pre- and post-design methodological issue to take into consideration—combining treated units that have been treated at different points in time. There are two potential problems for understanding use, as described earlier—if longitudinal data are used, then units are being treated at different ages and this can influence how they respond to legalization, and markets take time to develop, so factors that influence use (e.g., quality-adjusted prices) may differ as time elapses since passing a law. These challenges can be handled through model specification and estimation procedures during an analysis. If the issue is missing data throughout the pre-and/or post-period, there are a number of strategies one can use [for examples with drug survey data, see Miles and Hunt (2015)].

3.3 *What to Do When Randomly Assigning Laws to Jurisdictions Is not Feasible*

The third aspect of experimental designs is randomization, or the process of randomly assigning units to the control and treatment groups. The purpose of randomization, as described earlier, is to ensure that the treatment and control groups are equivalent (at the population level) so that all differences are due to chance rather than systematic variation. If the control group and treatment group are not equivalent, causal estimates are likely to be biased. For example, what we know from the descriptive data is that household survey respondents in states with MMLs are far more likely to report using marijuana in the last 12 months than those in non-MML states (Cerdá et al. 2012). But this does not necessarily mean having a MML *caused* MML use to increase. Maybe (probably) states with higher levels of marijuana use are also more likely to pass a MML. So a researcher needs to take into account the non-MML states (control group) are different from the MML states (treatment group) even before the introduction of the intervention.

Given the availability of survey data in the USA, researchers have been able to obtain pre- and post-intervention data and define a control group and a treatment group, but a method other than random assignment is used to assign units to the treatment and control groups. In this quasi-experimental situation, a statistical approach may be used to attempt to achieve unbiased causal inference. There are five key methods that researchers can consider where there is potential for non-equivalence of treatment and control groups:

Difference-in-difference (DiD) approach involves estimating the difference between outcome measures at two points (or more) for both the control group and the treated group and then comparing the pre–post differences between the groups. The idea is that any constant variable correlated with the selection decision and the outcome variable (e.g., cultural acceptance for marijuana) is 'differenced out' and does not bias the estimated effect of the law on the outcome—and that this is the

Fig. 1 Representation of DiD approach

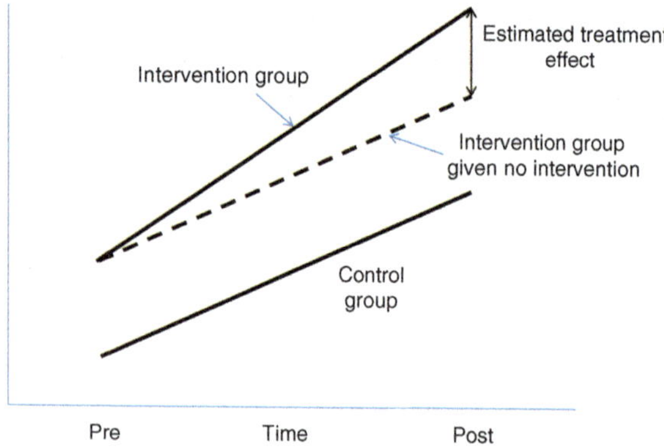

case regardless of whether it was measured or not. One key assumption of the approach is the parallel trends assumption in which the pre-period trends of the control group mirror those of the treatment group. The levels do not need to be the same because the approach takes into account differences between the groups, but the two groups need to be following the same trend before the intervention. Applications of this approach have been applied to estimate the impact of any MML on youth use (Harper et al. 2012; Anderson et al. 2015), any MML on accidents and alcohol use (Anderson et al. 2013), particular MML provisions on youth use and treatment episodes (Pacula et al. 2015), particular elements of MMLs on DSM-IV dependence and frequency of use in adults and youth (Wen et al. 2015), and particular MML provisions on potency (Sevigny et al. 2014). Figure 1 shows a simplified representation of the DiD approach. The control group is shown to be on an upward trajectory from pre- to post-intervention (e.g., increased marijuana use). The dashed line represents the change in the intervention group that would have been expected had the intervention not occurred—this is assumed to be parallel to the control group line. The upper solid line shows the actual change that occurred for the intervention group. The difference between the solid line and dashed line for the intervention group is the estimated treatment effect. For a helpful illustration of the control and intervention groups in a DiD design when units are treated at different times ('switching replications' design), see Lynne-Landsman et al. (2013).

A regression discontinuity (RD) design identifies the causal effects of an intervention by assigning a cutoff or threshold above or below which units are assigned to the control group or the treatment group. The idea is that units near the cutoff, yet on opposing sides, are similar to each other, but a cutoff has to be made somewhere. By comparing the outcomes of those meeting the criteria (treatment group) to the counterfactual outcome of those who just missed meeting the criteria (control group), an analysis estimates a local treatment effect. With respect to MMLs and RMLs, there is not an obvious RD design for passing the law or across provisions of the law. Indeed, we were not able to identify any research using an RD design to estimate the impact of changes in these laws. An example from alcohol policy research is to analyze the effects of the minimum legal drinking age of 21 on alcohol and marijuana use (Crost and Guerrero 2012; Crost and Rees 2013; Yörük and Yörük 2013). Given this other body of research, if the cutoff age for accessing alcohol is different than that for marijuana, there may be scope for considering how age cutoffs for accessing marijuana dispensaries affect use. It is important to ensure this key assumption is not violated when conducting causal analysis using an RD approach—another treatment cannot occur for the same cutoff.

Propensity score (PS) analysis incorporates a family of statistical techniques designed to use information to balance the baseline covariate distributions between treatment and control groups to ensure that the two groups are equal on these covariates. A propensity score is estimated, which is the conditional probability of treatment given the covariates of the unit of analysis, and the score is used in a model to estimate the impact of the intervention (McCaffrey et al. 2004). Table 1 shows a sample of pretreatment characteristics between the intervention and control groups before and after using a propensity score weighting technique in McCaffrey et al. (2004). The table shows the mean and standard deviation (*M, SD*) for each group, and the effect sizes (*d*) are calculated as the difference between group means divided by the standard deviation for the treatment group.[5] The left-hand side of the table (*Unweighted* columns) shows the makeup of the intervention and control groups prior to weighting. The groups are clearly unbalanced with regard to the covariates, with a total average absolute effect size of 0.31. The propensity score procedure balances the intervention and control groups, as shown on the right-hand side of the table (*Propensity Score Weighted* column). The average absolute effect size difference between the groups is reduced to 0.11. Similar to the RD approach, we were not able to identify any MML or RML impact studies using propensity score approaches. In part, this may be due to the fact that many studies use state representative data, so the control and treated groups are already balanced on observable characteristics.

The synthetic control (SC) approach is related to the propensity score and is used in settings where a single unit (a county, state, country, etc.) is exposed to an event or intervention and there are many untreated units for comparison. SC is a data-driven procedure in which a synthetic control unit is constructed based on a weighted combination of units in a comparison group that approximates the characteristics of treatment unit (Abadie et al. 2010). The basic idea is that not all control units are useful, but there may be some weighted combination of them that provides a better comparison for the treatment unit. Once again, we were not able to locate any published, peer-reviewed articles that have used a synthetic control approach to examine the effects of MMLs and RMLs. Given multiple jurisdictions have passed MMLs and now RMLs, one reason may be that researchers are not analyzing one treatment unit. However, using a SC approach for each of the states that have passed an MML or RML could be useful in substantiating results given these methods only reduce the bias of non-randomization, they do not eliminate it (Manski 2007).

---

[5]"The standard deviation for the treatment group is unaffected by propensity score weighting and allows for comparison pre- and post-weighting."

Table 1 Illustration of a PS approach, from McCaffrey et al. (2004)

	Unweighted					Propensity score weighted[a]			
	Intervention group		Control group			Control group			
Covariate	*M*	*SD*	*M*	*SD*	*d*	*M*	*SD*	*d*	*p*
Demographics									
Age (years)	15.82	0.91	15.31	1.28	0.56	15.76	1.11	0.07	0.58
Race (%)									
African American	8.57	28.07	18.61	38.99	–0.36	12.23	32.76	–0.13	0.26
Latino/Hispanic	60.00	49.13	52.19	50.04	0.16	55.24	49.73	0.1	0.43
White	20.57	40.54	13.14	33.84	0.18	18.62	38.92	0.05	0.72
Female (%)	18.29	38.77	9.12	28.85	0.24	8.17	27.39	0.26	0.01
School/work participation									
Last grade completed	9.04	1.25	8.59	1.36	0.36	8.85	1.28	0.15	0.20
Recency of last school/training[b]	5.41	1.16	5.28	1.36	0.11	5.22	1.41	0.16	0.21
Recency of paid work[b]	1.32	1.45	1.13	1.32	0.13	1.16	1.35	0.11	0.34
Current drug use/problems									
Days of alcohol/drug use (in past 90 days)[c]	6.19	3.17	3.77	3.59	0.76	5.59	3.35	0.19	0.1
Days drunk/high (in past 90 days)[c]	4.18	3.43	2.51	3.17	0.49	3.91	3.38	0.08	0.51
Substance involvement[c]	0.86	0.81	0.4	0.66	0.56	0.77	0.82	0.1	0.44
Substance-use intensity index[c]	7.61	4.36	4.59	4.72	0.69	6.94	4.75	0.16	0.19
Substance problem index (in past month)[c]	1.61	1.32	0.7	1.08	0.68	1.39	1.28	0.16	0.17
Current withdrawal index	1.53	0.28	1.49	0.24	0.16	1.51	0.26	0.08	0.47
[*rest of covariates not shown here*] …	…	…	…	…	…	…	…	…	…
Number of observations	75	274							
Effective number of observations						107.5			
Average absolute effect size					0.31			0.11	

Source McCaffrey et al. (2004)

[a]Treatment cases are not weighted, so only values for the control cases change with weighting

[b]Recency scale spans 0 (never) to 6 (past two days)

[c]Past 90-day frequency and count variables with a range greater than 15 are square-root-transformed to reduce variable skew

The instrumental variable (IV) method involves the identification of a variable (the instrument) that is directly correlated with the variable of interest (e.g., passing a MML), but is not directly correlated with the outcome of interest (e.g., marijuana use). In practice, the IV method is a two-step procedure in which the instrument is used to generate a predicted value of the independent variable (stage one), which is then used in the main model to estimate the impact of the independent variable on the outcome (stage two). It is typically very challenging to identify a valid instrument, i.e., one that is not directly correlated with the outcome of interest, so it is perhaps not surprising that once again, we were not able to identify any studies using an IV approach to estimate the impact of MMLs or RMLs. One general example of an IV estimation in the context of marijuana policy analysis is Desimone (1998). The aim of the study was to estimate the effect of past marijuana use on cocaine use, e.g., whether marijuana led to cocaine use, but these two variables are spuriously correlated because unobserved factors (e.g., tastes for intoxication, deviance, lower risk aversion) may simultaneously affect the use of marijuana and cocaine. In other words, people were not randomly assigned to consume marijuana. The author used three instruments affecting past marijuana demand, but not current cocaine demand—marijuana penalties, beer taxes, and alcoholic parents—and found using marijuana increases the probability of using cocaine later.

4 Emerging Empirical Findings

There is a growing body of research focused on estimating the causal impact of marijuana laws on consumption or use of marijuana. Table 2 presents a summary of studies estimating the impact of MMLs on youth and young adult use using statistical approaches for quasi-experimental settings. Although the studies use differing data sources, time frames, assignment of states to control and treatment groups, coverage, and model specifications (although they all apply DiD), they all find no statistically significant effect of a MML as a dichotomous indicator on past month use or frequency of use for youth. There are two statistically significant findings for the treatment and initiation; having any law in place reduces treatment episodes where the primary substance of admission is marijuana (Pacula et al. 2015) and increases the probability of first-time marijuana use between the ages of 12 and 20 (Wen et al. 2015). It is unclear whether the treatment episode reduction result is due to less reporting of marijuana as the primary problem or an actual reduction in dependence/abuse of marijuana; using a different measure of dependence (DSM-IV), no change was found as a result of passing any MML (Wen et al. 2015). Combining these results suggests having a MML increased the likelihood of young people trying marijuana, but they use relatively infrequently because there was no change in past month prevalence or high frequency use (20 or more days in last month).

Table 2 Summary of the literature estimating the impact of MMLs on youth use

	Time frame	Data source	Intervention definition	Number of treatment and control groups	Outcome measured	Finding
Under 18 years of age						
Anderson et al. (2015)	1993–2011 Biannual, some gaps in data (not all years for every state)	YRBS, national and state representative, 9th–12th graders*, individual data	Any MML	T: 16 states C: the other states in the group of 16 that had not yet passed the MML and the 34 other states	Past month Frequent use Use at school Offered or bought at school	NS NS NS NS
Lynne-Landsman et al. (2013)	2003–2011 Biannual, some gaps in data (not all years for every state)	YRBS, 9th–12th grade*, individual data	Any MML	T: 4 states C: the other states in the group of 4 that had not yet passed the MML	Past month Frequent use	NS NS
Harper et al. (2012)	2002–2009 Annual	NSDUH, 12–17 year olds, state-level data	Any MML	T: 5 states adopt MML during period and 8 states with MML entire period C: 3 states with a MML after 2009 and 25 states without a MML prior to 2013	Past month use Perceived riskiness of monthly use	NS NS
Gorman and Huber (2007)	1995–2002 Quarterly	ADAM, 10–18 year olds, individual data		T: 4 cities (3 in CA, 1 in OR) C: none (interrupted time series design)	Proportion testing positive	Results for CA law/OR law outcome: NS/NS

(continued)

Table 2 (continued)

	Time frame	Data source	Intervention definition	Number of treatment and control groups	Outcome measured	Finding
Under 21 years of age						
Wen et al. (2015)	2004–2012 Monthly	NSDUH, 12–20 year olds, individual data	Any MML Dispensary Home cultivation Patient registry Non-specific pain	T: 10 states C: 8 states with a MML prior to 2004 and 31 states without a MML prior to 2012 (for T and C by law type, see Wen et al. (2015))	Past month use Frequent use Initiation Abuse/dependence	Results for any/dispensary/home/registry/pain by outcome: NS/Positive/NS/NS/NS NS/NS/NS/NS Increase/NS/NS/NS NS/NS/NS/NS
Pacula et al. (2015)	1992–2011 Annual 1997–2005 Annual	TEDS, 12–20 year olds, individual data NLSY97, 12–20 year olds, individual data	Any MML Dispensary Home cultivation Patient registry	For TEDS, T: possibly 15 states adopt MML during period (and potentially 3 states in the final year of data) C: possibly the other states in the group of 18 that had not yet passed the MML and possibly the other 23 states For NLSY97, T: possibly 8 states adopt MML during period (and potentially 3 states with MML in the first or final year of data) C: possibly the other states in the group of 11 that had not yet passed the MML and possibly the other 30 states	With TEDS: MJ as primary substance treatment admission With NLSY97: Past month use Frequent use # days use past month	Results for any/dispensary/home/registry by outcome: Decrease/increase/NS/NS NS/NS/negative/positive NS/NS/negative/positive NS/NS/negative/positive

(continued)

Table 2 (continued)

	Time frame	Data source	Intervention definition	Number of treatment and control groups	Outcome measured	Finding
Young Adults						
Harper et al. (2012)	2002–2009 Annual	NSDUH, 18–25 year olds, state-level data	Any MML	T: 5 states adopt MML during period and 8 states with MML entire period C: 3 states with a MML after 2009 and 25 states without a MML prior to 2013	Past month use Perceived riskiness of monthly use	NS NS
Gorman and Huber (2007)	1995–2002 Quarterly	ADAM, 18–20 year olds, individual data		With ADAM, T: 5 cities (3 in CA, 1 in CO, 1 in OR) C: none (interrupted time series design)	Proportion testing positive	Results for CA law/CO law/OR law by outcome: NS/NS/NS

*Range 12–18 year olds
NS Not statistically significant. ‘States’ include Washington D.C.

There are some differences when the intervention definition refers to passing particular provisions in the law, rather than generally passing any law (for a description on intervention definition issues, see the previous section). Legalizing dispensaries increases the proportion of young people who used in the past month, but again does not affect the frequency of use or dependence (Wen et al. 2015). One study finds allowing home cultivation reduces use and requiring patient registries increases use (Pacula et al. 2015), while another study finds these laws do not have a statistically significant impact (Wen et al. 2015). There are a number of differences between these two studies that may, or may not, drive these differential findings: Pacula et al. (2015) focus on 1997 through 2005 with a control group of individuals in non-MML (yet) states, whereas Wen et al. (2015) examine the dispensary laws after this period from 2004 to 2012 with a control group of individuals in non-MML states and early adopter MML states. Therefore, it could be that the small impacts of Pacula et al. (2015) are attenuated when the control group includes states that have already passed MMLs and/or that the effects in the earlier states differed from those adopting later. Furthermore, Wen et al. (2015) use a repeated cross section, whereas individuals are aging out of the dataset in Pacula et al. (2015), so the timing of the passage of the law matters in that study.

For young adults, the two studies use different data sources, time frames, assignment of states to control and treatment groups, coverage, and research designs [DiD (Harper et al. 2012) and interrupted time series (Gorman and Huber 2007)] and find no statistically significant effect of a MML as a dichotomous indicator on use, perceived riskiness of use, or use during arrest.

Table 3 presents a summary of studies for adult use. For an intervention defined as having any MML, there appears to be an increase in the past month and highly frequent use for adults aged 21 and over and treatment for males over 18 years of age, yet no statistically significant impact on dependence/abuse. Since this intervention definition allows for heterogeneous treatment (i.e., provisions in the laws vary across states), research examines specific laws and finds this result may be driven by allowing physicians to recommend medical marijuana for pain without having to identify the condition or cause of pain. When the intervention refers to laws of five particular states, there is no statistically significant impact at an aggregate level for adults aged 26 and over, and no impact on emergency room admissions mentioning marijuana in California, Colorado, or Washington.

When examining all ages, 10 or 12 years of age and older, a slightly different picture emerges. For an intervention defined as having any MML, there is no statistically significant impact on past month use or frequency of use, compared to the positive impact found for 21+ year olds, and no impact on arrestees testing positive for marijuana. When isolating the intervention definition to specific provisions in the laws, results indicate that allowing home cultivation increased all measures of use, dispensaries increased past month use, patient registries have no effect, and all other features decreased past month and frequent use for those aged 12 and older.

Table 3 Impacts of MMLs on adult use in the previous literature

	Time	Data	Intervention definition	Control and treatment groups	Outcome	Finding
Adult population						
Wen et al. (2015)	2004–2012	NSDUH, 21 + year olds, individual data	Any MML Dispensary Home cultivation Patient registry Non-specific pain	T: 10 states C: 8 states with a MML prior to 2004 and 31 states without a MML prior to 2012 T: 9 states T: 13 states T: 14 states T: 14 states	Past month use Frequent use Initiation Abuse/dependence	Results for any/dispensary/home/registry/pain by outcome: +/NS/NS/NS/+ +/NS/NS/NS/+ NS/NS/NS/NS NS/NS/NS/NS
Chu et al. (2014)	1992–2008	TEDS, 18+ year olds/adult males, non-CJ referrals, individual data	Any MML CA law specifically CO law specifically	For any MML: T: 12 states C: 10 states with a MML after 2008 and the other 29 states	Marijuana-related treatment ratio (to all substances) Ratio of first-time marijuana treatment (to all times)	Results for any/CA law/CO law by outcome: +/NS/- +/NS/NS
Harper et al. (2012)	2002–2009 Annual	NSDUH, 26 + year olds, state-level data	Any MML	T: 5 states adopt MML during period and 8 states with MML entire period C: 3 states with a MML after 2009 and 25 states without a MML prior to 2013	Past month use Perceived riskiness of monthly use	NS NS
Gorman and Huber (2007)	1994–2002 Quarterly	DAWN, 18 + year olds, individual data	Any MML	T: 5 cities (3 in CA, 1 in CO, 1 in WA) C: none (interrupted time series design)	Proportion of emergency department visits mentioning marijuana	Results for CA law/CO law/WA: NS/NS/NS

(continued)

Table 3 (continued)

	Time	Data	Intervention definition	Control and treatment groups	Outcome	Finding
All ages						
Pacula et al. (2015)	1992–2011 Annual 1997–2011 Annual	TEDS, 12+ year olds, individual data NLSY, 12+ year olds, individual data	Any MML Dispensary Home cultivation Patient registry	For TEDS, T: possibly 15 states adopt MML during period (and potentially 3 states in the final year of data) C: possibly the other states in the group of 18 that had not yet passed the MML and possibly the other 23 states For NLSY97, T: possibly 8 states adopt MML during period (and potentially 3 states with MML in the first or final year of data) C: possibly the other states in the group of 11 that had not yet passed the MML and possibly the other 30 states	With TEDS: MJ as primary substance treatment admission With NLSY97: Past month use Frequent use # days use past month	Any/dispensary/home/registry by outcome: −/−/NS/NS NS/+/+/NS NS/NS/+/NS NS/NS/+/NS
Gorman and Huber (2007)	1995–2002 Quarterly	ADAM, 10 + year olds, individual data	Any MML	T: 5 cities (3 in CA, 1 in CO, 1 in OR) C: none (interrupted time series design)	Proportion testing positive	Results for CA law/CO law/OR law: NS/NS/NS

Pacula et al. (2015) find for a decrease in treatment episodes with marijuana as the primary substance of abuse for individuals aged 12 and older, which appears to be driven by state laws allowing for dispensaries, compared to a positive impact on the ratio of marijuana treatment episodes to all other substances' treatment for individuals aged 18 and older. Given the two measures used in these studies, it is possible the number of treatment episodes for all substances fell during the period investigated (number of marijuana episodes decreased), but marijuana treatments fell by less than treatment for other substances (marijuana as a ratio increased); for more on measurement issues, see the previous section.

While the studies analyzing different ages and time periods find differential effects on use, perhaps more importantly, they also use different intervention definitions. Pacula et al. (2015) appear to define a state as treated if evidence suggests the particular law was implemented, whereas Wen et al. (2015) appear to define a state as treated if a medical-related law was passed and the provision of interest, e.g., dispensaries, was not explicitly prohibited. For example, the first law passed in CA in 1996 (Proposition 215) was silent on the issue of dispensaries, but in 2003, a state statute (SB 420) was passed explicitly permitted dispensaries (Pacula et al. 2014) and evidence suggests dispensaries were operating in 2003 (Pacula et al. 2015). CA is considered treated in 1996 in Wen et al. (2015) and treated in 2003 in Pacula et al. (2015).

4.1 Impact of Legalizing a Recreational Market for Marijuana

The reality about understanding the effect of a recreational market on health and safety outcomes is simple—we do not have answers yet. At the time this study was submitted, a fully operating, legal market for recreational marijuana has only existed in Colorado for 1.5 years and in Washington for 1 year. Even if a researcher is analyzing the short-term effects of legalization, it takes time to prepare the data and identify the appropriate research design, conduct the analysis, present research to the academic community for feedback, edit the analysis and/or paper, submit the final paper to a journal, allow time for reviewers to generate their referee reports, and then revise the manuscript and resubmit to the journal for copyediting and publication. Given the number of working papers out now, several peer-reviewed analyses on the short-term impacts of opening a recreational market on use, crime, and affordability of marijuana should be published soon. This is not to be confused with the long-run or steady-state impact evaluations for which policymakers are really waiting. Those studies require several more years of data to allow the behavior of suppliers and consumers to respond to the change in the law, so it will take some time before studies with strong research designs can provide those answers.

5 Conclusion

Advancements in statistical approaches to quasi-experimental settings have opened doors for researchers to address pressing questions concerning policy impacts when experimentation is simply not possible. This paper focuses on three main elements of research design—defining the intervention, measurement, and randomization—as they relate to studying the impact of MMLs and RMLs on use. We show that results vary across studies because they measure use differently and define the intervention differently. Some studies research the question of whether passing any law on medical marijuana has an impact on use. Other studies focus on whether passing a particular provision in the law, e.g., permitting dispensaries, has an impact on use. Furthermore, some studies research whether *explicitly* passing a law has an impact on use and others research whether a provision has an impact on use if a law was passed, but a particular provision was not explicitly mentioned or clear. The aim of this paper is not to recommend a particular definition, but rather to show that the definition affects results and demonstrate that the intervention definition should be directly linked to the research question.

This study also synthesizes literature findings of studies using causal inference approaches to estimate the impact of MMLs and RMLs on youth and adult use. The research tends to show that having any MML increases the likelihood of youth trying marijuana, but no statistically significant effects on frequency of use in terms of past month prevalence or heavy use (20 or more days in last month) have been found. Decomposing the law into particular provisions that likely increase the availability and accessibility of marijuana (e.g., home cultivation, dispensary, patient registration, and pain definitions) results in differential findings across studies on past month use and frequency of use. Similarly, the impact of MMLs on adult use is complex because studies find differential impacts. We show that these seemingly conflicting findings may be due to different years of analysis, ages of the sample in each year, and assignment of states to the control group versus treatment group. The predominant statistical approach thus far to estimate the causal effect of MMLs on use is differences-in-differences, and we discuss a number of design reasons why this might be. It will be important in the future to understand whether other approaches, such as propensity score and synthetic control methods, generate similar results.

There are a number of other issues of interest to policymakers and researchers that this paper does not address. Given opioid overdose rates in the USA and an association between a MML and opioid overdose (Bachhuber et al. 2014), it would be important to understand whether this relationship is indeed causal. Furthermore, studies show consumption relationships between marijuana and alcohol (Midanik et al. 2007; Pacula 1998) and marijuana use and motor vehicle accidents (Li et al. 2012), so there are questions about whether MMLs and RMLs affect fatal traffic accidents (Anderson et al. 2013) and non-fatal accidents. A potentially important mechanism to better understand the effect of MMLs and RMLs on public safety and health outcomes is product development and potency. Even if the laws do not result in people consuming more or more people consuming, MMLs and RMLs could have health and safety impacts because the law results in stronger marijuana (Sevigny et al. 2014). Another imperative avenue of research would be to investigate how the laws may affect crime and the criminal justice system. Studies are starting to examine effects on arrests for marijuana (Chu 2014), and we should start to see analyses of violence in the black market, property and violent crime directed at legal suppliers (because of a restriction on credit card transactions, for example), and criminal behavior and victimization of users.

Acknowledgements The authors would like to acknowledge funding from the National Institute of Drug Abuse, grants #R01DA032693 and #R01DA032693-03S1.

References

Abadie A, Diamond A, Hainmueller J (2010) Synthetic control methods for comparative case studies: estimating the effect of California's tobacco control program. J Am Stat Assoc 105:493–505. doi:10.1198/jasa.2009.ap08746
Anderson D, Hansen B, Rees DI (2013) Medical marijuana laws, traffic fatalities, and alcohol consumption. J Law Econ 56:333–369. doi:10.1086/668812
Anderson D, Hansen B, Rees DI (2015) Medical marijuana laws and teen marijuana use. Am Law Econ Rev. doi:10.1093/aler/ahv002
Anderson DM, Rees DI (2014) The role of dispensaries: the devil is in the details: point/counterpoint. J Policy Anal Manage 33:235–240. doi:10.1002/pam.21733
Angrist JD, Imbens GW, Rubin DB (1996) Identification of causal effects using instrumental variables. J Am Stat Assoc 91:444–455. doi:10.1080/01621459.1996.10476902
Ashenfelter O, Card D (1985) Using the longitudinal structure of earnings to estimate the effect of training programs. Rev Econ Stat 67:648. doi:10.2307/1924810
Bachhuber MA, Saloner B, Cunningham CO, Barry CL (2014) Medical cannabis laws and opioid analgesic overdose mortality in the United States, 1999–2010. JAMA Intern Med 174:1668. doi:10.1001/jamainternmed.2014.4005
Campbell DT, Stanley JC (1963) Experimental and quasi-experimental Designs for Research. Boston: Houghton Mifflin Company
Cerdá M, Wall M, Keyes KM et al (2012) Medical marijuana laws in 50 states: investigating the relationship between state legalization of medical marijuana and marijuana use, abuse and dependence. Drug Alcohol Depend 120:22–27. doi:10.1016/j.drugalcdep.2011.06.011
Chu Y-WL (2014) The effects of medical marijuana laws on illegal marijuana use. J Health Econ 38:43–61. doi:10.1016/j.jhealeco.2014.07.003
CODoR (2014) State of Colorado Marijuana Taxes, licenses, and fees transfers and distribution: May 2014 sales reported in June. Colorado Department of Revenue
Cohen JS (1988) Statistical power analysis for the behavioral sciences, 2nd edn. Erlbaum, Hillsdale
Crost B, Guerrero S (2012) The effect of alcohol availability on marijuana use: evidence from the minimum legal drinking age. J Health Econ 31:112–121. doi:10.1016/j.jhealeco.2011.12.005
Crost B, Rees DI (2013) The minimum legal drinking age and marijuana use: new estimates from the NLSY97. J Health Econ 32:474–476. doi:10.1016/j.jhealeco.2012.09.008
DeSimone J (1998) Is marijuana a gateway drug? Eastern Econ J 24:149–164. doi:10.2307/40325834
Fisher RA (1954) Statistical Methods for Research Workers (12 ed.). Edinburgh: Oliver and Boyd

Gorman DM, Huber J (2007) Do medical cannabis laws encourage cannabis use? Int J Drug Policy 18:160–167. doi:10.1016/j.drugpo.2006.10.001
Guilford JP (1942) Fundamental Statistics in Psychology and Education. New York: McGraw-Hill
Harper S, Strumpf EC, Kaufman JS (2012) Do medical marijuana laws increase marijuana use? Replication study and extension. Ann Epidemiol 22:207–212
Holland PW (1986) Statistics and causal inference. J Am Stat Assoc 81:945–960.
Kenny DA (1979) Correlation and Causality. New York: Wiley
Kilmer B, Everingham S, Caulkins J et al (2014) What America's users spend on illegal drugs: 2000–2010. Office of National Drug Control Policy, Washington
Lee DS, Lemieux T (2010) Regression discontinuity designs in economics. J Econ Lit 48:281–355. doi:10.1257/jel.48.2.281
Li M-C, Brady JE, DiMaggio CJ et al (2012) Marijuana use and motor vehicle crashes. Epidemiol Rev 34:65–72. doi:10.1093/epirev/mxr017
Lynne-Landsman SD, Livingston MD, Wagenaar AC (2013) Effects of state medical marijuana laws on adolescent marijuana use. Am J Public Health 103:1500–1506. doi:10.2105/AJPH. 2012.301117
Manski C (2007) Identification for prediction and decision. President and Fellow of Harvard College, Cambridge
McCaffrey DF, Ridgeway G, Morral AR (2004) Propensity score estimation with boosted regression for evaluating causal effects in observational studies. Psychol Methods 9:403–425. doi:10.1037/1082-989X.9.4.403
Medical Research Council (1948) Streptomycin treatment of pulmonary tuberculosis: a report of the Streptomycin in Tuberculosis Trials Committee. Brit Med J 40(4882):769–781
Midanik LT, Tam TW, Weisner C (2007) Concurrent and simultaneous drug and alcohol use: results of the 2000 national alcohol survey. Drug Alcohol Depend 90:72–80. doi:10.1016/j. drugalcdep.2007.02.024
Miles J, Hunt P (2015) A practical introduction to methods for analyzing longitudinal data in the presence of missing data using a marijuana price survey. J Crim Psychol 5(2)
MPP (2013) Amendment 64: an overview of Colorado's new approach to marijuana. Marijuana Policy Project
NCSL (2013) Oversight of the U.S. drug supply: H.R. 3204, the drug quality and security act. National Conference of State Legislatures
Pacula RL (1998) Does increasing the beer tax reduce marijuana consumption? J Health Econ 17:557–585. doi:10.1016/S0167-6296(97)00039-8
Pacula RL, Boustead A, Hunt P (2014) Words can be deceiving: a review of variation among legally effective medical marijuana laws in the United States. J Drug Policy Anal 7(1):1–19
Pacula RL, Heaton P, Powell D, Sevigny E (2015) Assessing the effects of medical marijuana laws on marijuana and alcohol use: the devil is in the details. J Policy Anal Manage. doi:10.1002/ pam.21804
Pacula RL, MacCoun R, Reuter P et al (2005) What does it mean to decriminalize marijuana? A cross-national empirical examination. Adv Health Econ Health Serv Res 16:347–369
Pearl J (2000) Causality: models, inference and reasoning. Cambridge: Cambridge University Press.
Room R (2014) Legalizing a market for cannabis for pleasure: Colorado, Washington, Uruguay and beyond: legalising a market for cannabis. Addiction 109:345–351. doi:10.1111/add.12355
Rosenbaum PR, Rubin DB (1983) The central role of the propensity score in observational studies for causal effects. Biometrika 70:41–55. doi:10.1093/biomet/70.1.41
Rubin DB (1974) Estimating causal effects of treatments in randomized and nonrandomized studies. J Educ Psychol 66:688–701. doi:10.1037/h0037350
Russell B (1912) On the Notion of Cause. Proceedings of the Aristotelian Society 13:1–26
Sevigny EL (2013) Is today's marijuana more potent simply because it's fresher? Marijuana potency. Drug Test Anal 5:62–67. doi:10.1002/dta.1430
Sevigny EL, Pacula RL, Heaton P (2014) The effects of medical marijuana laws on potency. Int J Drug Policy 25:308–319. doi:10.1016/j.drugpo.2014.01.003

Wen H, Hockenberry JM, Cummings JR (2015) The effect of medical marijuana laws on adolescent and adult use of marijuana, alcohol, and other substances. J Health Econ 42:64–80. doi:10.1016/j.jhealeco.2015.03.007

Yörük BK, Yörük CE (2013) The impact of minimum legal drinking age laws on alcohol consumption, smoking, and marijuana use revisited. J Health Econ 32:477–479. doi:10.1016/j.jhealeco.2012.09.007

Zhang Z (2003) Drug and alcohol use and related matters among arrestees 2003. National Opinion Research Center, Washington, D.C.

CPSIA information can be obtained
at www.ICGtesting.com
Printed in the USA
LVOW05*2109210218
567424LV00001B/71/P

9 783319 600147